# Assessing and Managing the Ecological Impacts of Paved Roads

Committee on Ecological Impacts of Road Density

Board on Environmental Studies and Toxicology

Division on Earth and Life Studies

Transportation Research Board

NATIONAL RESEARCH COUNCIL
*OF THE NATIONAL ACADEMIES*

THE NATIONAL ACADEMIES PRESS
Washington, D.C.
**www.nap.edu**

**THE NATIONAL ACADEMIES PRESS**   500 Fifth Street, NW   Washington, DC 20001

NOTICE: The project that is the subject of this report was approved by the Governing Board of the National Research Council, whose members are drawn from the councils of the National Academy of Sciences, the National Academy of Engineering, and the Institute of Medicine. The members of the committee responsible for the report were chosen for their special competences and with regard for appropriate balance.

This project was supported by Contract No. DTFH61-01-C-00036 between the National Academy of Sciences and the Department of Transportation. Any opinions, findings, conclusions, or recommendations expressed in this publication are those of the author(s) and do not necessarily reflect the view of the organizations or agencies that provided support for this project.

Cover design by Liza R. Hamilton, National Research Council. Front photo by Anthony P. Clevenger, Montana State University. Back photos used with permission from Lance H. Gunderson, Emory University (double yellow line), Emory University Facilities Management (campus), and USGS (southeastern United States).

International Standard Book Number 0-309-10088-7 (Book)
International Standard Book Number 0-309-65631-1 (PDF)

Library of Congress Control Number 2005937774

Additional copies of this report are available from

The National Academies Press
500 Fifth Street, NW
Box 285
Washington, DC 20055

800-624-6242
202-334-3313 (in the Washington metropolitan area)
http://www.nap.edu

# THE NATIONAL ACADEMIES
## Advisers to the Nation on Science, Engineering, and Medicine

The **National Academy of Sciences** is a private, nonprofit, self-perpetuating society of distinguished scholars engaged in scientific and engineering research, dedicated to the furtherance of science and technology and to their use for the general welfare. Upon the authority of the charter granted to it by the Congress in 1863, the Academy has a mandate that requires it to advise the federal government on scientific and technical matters. Dr. Ralph J. Cicerone is president of the National Academy of Sciences.

The **National Academy of Engineering** was established in 1964, under the charter of the National Academy of Sciences, as a parallel organization of outstanding engineers. It is autonomous in its administration and in the selection of its members, sharing with the National Academy of Sciences the responsibility for advising the federal government. The National Academy of Engineering also sponsors engineering programs aimed at meeting national needs, encourages education and research, and recognizes the superior achievements of engineers. Dr. Wm. A. Wulf is president of the National Academy of Engineering.

The **Institute of Medicine** was established in 1970 by the National Academy of Sciences to secure the services of eminent members of appropriate professions in the examination of policy matters pertaining to the health of the public. The Institute acts under the responsibility given to the National Academy of Sciences by its congressional charter to be an adviser to the federal government and, upon its own initiative, to identify issues of medical care, research, and education. Dr. Harvey V. Fineberg is president of the Institute of Medicine.

The **National Research Council** was organized by the National Academy of Sciences in 1916 to associate the broad community of science and technology with the Academy's purposes of furthering knowledge and advising the federal government. Functioning in accordance with general policies determined by the Academy, the Council has become the principal operating agency of both the National Academy of Sciences and the National Academy of Engineering in providing services to the government, the public, and the scientific and engineering communities. The Council is administered jointly by both Academies and the Institute of Medicine. Dr. Ralph J. Cicerone and Dr. Wm. A. Wulf are chair and vice chair, respectively, of the National Research Council.

**www.national-academies.org**

# ACKNOWLEDGMENTS

We are grateful for the generous support provided by U.S. Department of Transportation and are especially grateful for the outstanding assistance provided by Dr. Paul Garrett. Many people assisted the committee and National Research Council by providing data and reports. We are grateful for the information and support provided by the following:

Marina Alberti, University of Washington
David Andersen, Washington State Office of Community Trade and Economic Development
Fred Bank, Federal Highway Administration
Paul Barten, University of Massachusetts
William I. Boarman, U.S. Geological Survey
Ray Bransfield, U.S. Fish and Wildlife Service
Todd D. Carlson, Washington State Department of Transportation
Laurie W. Carr, TerraSystems Research
Kristan Majors Chilcoat, Emory University
Scott Dawson, California Department of Fish and Game
Robert Fuerstenberg, King County Department of Natural Resources and Parks
Margot J. Griswold, Earthworks Construction & Design
Howard Haemmerle, King County Department of Transportation
Eugene S. Helfman, University of Georgia
Art Homrighausen, LSA Associates, Inc.
Geary Hund, California State Parks
Robert A. James, California Department of Transportation
John Kusler, Association of State Wetland Managers
Gino Lucchetti, King County Department of Natural Resources
Gary McVoy, New York State Department of Transportation
Ivan Miller, Puget Sound Regional Council
William Ruediger, U.S. Forest Service
Mark Shaffer, Doris Duke Foundation
Emily Tibbott, The Nature Conservancy
Benjamin Tuggle, U.S. Fish and Wildlife Service
Sylvia Vega, California Department of Transportation
Paul A. Waddel, University of Washington
Michael White, Conservation Biology Institute

# Acknowledgment of Review Participants

This report has been reviewed in draft form by individuals chosen for their diverse perspectives and technical expertise, in accordance with procedures approved by the National Research Council's Report Review Committee. The purpose of this independent review is to provide candid and critical comments that will assist the institution in making its published report as sound as possible and to ensure that the report meets institutional standards for objectivity, evidence, and responsiveness to the study charge. The review comments and draft manuscript remain confidential to protect the integrity of the deliberative process. We wish to thank the following individuals for their review of this report:

Frank W. Davis, University of California, Santa Barbara
Andras Fekete, The RBA Group
Richard T.T. Forman, Harvard University
Kevin E. Heanue, Consultant
Arnold M. Howitt, Harvard University
Herbert S. Levinson, Consultant
Judith L. Meyer, University of Georgia
Helen Mountford, Consultant
G. Scott Rutherford, University of Washington
William H. Schlesinger, Duke University
Kumares C. Sinha, Purdue University
B.L. Turner II, Clark University
Susan L. Ustin, University of California, Davis
Patricia White, Defenders of Wildlife

Although the reviewers listed above have provided many constructive comments and suggestions, they were not asked to endorse the conclusions or recommendations, nor did they see the final draft of the report before its release. The review of this report was overseen by William L. Chameides, Georgia Institute of Technology; Gordon Orians, University of Washington (Emeritus); and Lester A. Hoel, University of Virginia. Appointed by the National Research Council, they were responsible for making certain that an independent examination of this report was carried out in accordance with institutional procedures and that all review comments were carefully considered. Responsibility for the final content of this report rests entirely with the authoring committee and the institution.

# Preface

In the past century, dramatic changes have been made in the U.S. road system to accommodate an evolving set of needs, including personal travel, economic development, and military transport. As the struggle to accommodate larger volumes of traffic continues, the road system is increasing in width and, at a slower pace, overall length.

As the road system changes, so does the relationship between roads and the environment. With the increase in roads, more resources are going toward road construction and management. More is also understood about the impact of roads on the environment. To address these matters, a better understanding of road ecology and better methods of integrating that understanding into all aspects of road development are needed.

This report attempts to consolidate understanding in a number of areas—how roads interact with different ecological structures and processes across scales of space and time; how the legal framework for evaluating ecological effects intersects with the scales of ecological features; and how ecological considerations can be integrated better into all phases of road development—from planning to use.

The compilation of current understanding of the effects of roads on ecological processes and structures is a major focus of this report. We found that most of the current knowledge is about aspects of the environment that change over short time periods and small areas and that ecological processes and structures that cover large areas over broad time scales have been largely overlooked.

The current federal legal framework for consideration of ecological impacts of roads has essentially been in place for over three decades.

This framework considers a few critical pieces of ecosystems but could be expanded to include more ecological features. The opportunity exists to evaluate the efficacy of these laws and policies, but the collection of more data and the synthesis of extant information need to be done in ways that have generally not been done.

Integrating ecological considerations into all phases of road development—from planning to construction to vehicle use to ongoing maintenance—is a continuing challenge. Practitioners are moving in that direction and are encouraged to continue in that direction. We suggest that integrative assessments done earlier in the planning process are a key solution to this chronic issue.

These findings would not have been possible without the hard work, collective action, and perseverance of this committee. I would like to thank my colleagues on the committee for their efforts. Some members of the group have hundreds of years of collective experience as practitioners and gladly shared that wisdom; others are among the leading researchers in the field of road ecology. I have been honored to serve alongside such an august group. But even more, it was a pleasure and joy to get to know them and work with them all.

The committee and I gratefully acknowledge and thank the staff of the National Research Council for their support. Suzanne van Drunick served as project director and provided gracious leadership throughout the project. Bryan Shipley provided beneficial research and report preparation. Liza Hamilton provided outstanding venues and programs for the meetings and help in preparing the report. David Policansky provided helpful guidance and counsel. Ruth Crossgrove, Mirsada Karalic-Loncarevic, John Brown, Alexandra Stupple, and Sammy Bardley also helped with various aspects of the project and report preparation. The committee is appreciative of BEST members for their oversight of this study.

The committee benefited immensely from the help and advice of practitioners and scientists who spent their valuable time to give presentations, reports, and advice to the committee during the numerous meetings.

The National Research Council process for producing the report involves extensive reliance on external reviewers. The committee thanks the reviewers of its final report for their thoughtful contributions.

We hope that the conclusions and recommendations of the report provide solid advice to agencies and the road community to better assess

and integrate environmental concerns into all phases of road development. We also hope that in some small way this information provides a
foundation for a more sustainable future.

*Lance Gunderson,* Chair
*Committee on Ecological*
*Impacts of Road Density*

# Abbreviations

| | |
|---|---|
| AASHTO | American Association of State Highway and Transportation Officials |
| BMP | best-management practices |
| CAA | Clean Air Act |
| Caltrans | California Department of Transportation |
| CE | categorical exclusion |
| CEQ | Council on Environmental Quality |
| CIFOR | Center for International Forestry Research |
| CMAQ | Congestion Mitigation and Air Quality Improvement Program |
| Corps | U.S. Army Corps of Engineers |
| CSD | context-sensitive design |
| CSS | context-sensitive solutions |
| CWA | Clean Water Act |
| CZMA | Coastal Zone Management Act |
| DOA | U.S. Department of Agriculture |
| DOI | U.S. Department of the Interior |
| DOT | U.S. Department of Transportation |
| EA | environmental assessment |
| EA/FONSI | environmental assessment and finding of no significant impact |
| EIS | environmental impact statement |
| EO | Executive Order |
| EPA | U.S. Environmental Protection Agency |
| EPS | environmental policy statement |
| ESA | Endangered Species Act |

| | |
|---|---|
| ETAP | environmental technical advisory team |
| FHWA | Federal Highway Administration |
| FLH | federal lands highway |
| FPPA | Farmland Protection Policy Act |
| FTA | Federal Transit Administration |
| GIS | geographic information system |
| HEP | habitat evaluation procedure |
| HGM | hydrogeomorphic method |
| IRI | international roughness index |
| ISTEA | Intermodal Surface Transportation Equity Act |
| LARCH | landscape ecological analysis and rules for the configuration of habitat model |
| LOS | level of service |
| LRTP | long-range transportation plan |
| LWCFA | Land and Water Conservation Fund Act |
| MBTA | Migratory Bird Treaty Act |
| MOU | memorandum of understanding |
| MPO | metropolitan planning organization |
| NAAQS | National Ambient Air Quality Standards |
| NAS | National Academy of Sciences |
| NCHRP | National Cooperative Highway Research Program |
| NEPA | National Environmental Policy Act |
| NGO | nongovernmental organization |
| NHPA | National Historic Preservation Act |
| NHS | national highway system |
| NOAA | National Oceanic and Atmospheric Administration |
| $NO_X$ | nitrogen oxides |
| NPDES | national pollutant discharge elimination system |
| NPS | National Park Service |
| NRC | National Research Council |
| NWP | nationwide permit |
| $PM_{10}$ | particulate matter up to 10 micrometers in diameter |
| ROD | record of decision |
| SAMP | special area management plan |
| SEA | street edge alternatives |
| SEF | southeastern ecological framework |
| SEMP | Strategic Environmental Management Program |
| SIP | state implementation plan |
| SPGP | state program general permit |
| STIP | state transportation improvement plan |

| STPP | surface transportation policy project |
| TE | transportation enhancement |
| TEA-21 | Transportation Equity Act for the Twenty-First Century |
| TIP | transportation improvement program |
| TMDL | total maximum daily load |
| TNC | The Nature Conservancy |
| TRB | Transportation Research Board |
| USFWS | U.S. Fish and Wildlife Service |
| V/SF | volume to service flow ratio |
| VMT | vehicle miles of travel |
| VOC | volatile organic compound |

# Contents

**APPENDIXES**

**FIGURES**

**TABLES**

# Assessing and Managing the Ecological Impacts of Paved Roads

# Summary

There are 4 million miles of roads in the United States. One hundred years ago, roads were primarily unpaved and had half the number of miles of the present U.S. road system. As the system grew, roads became wider and more complex structurally to provide for more and heavier traffic. New construction technology and greater structural stability were needed to improve the road system.

All phases of road development—from construction and use by vehicles to maintenance—affect physical and chemical soil conditions, water flow, and air and water quality. Roads alter habitats, increase wildlife mortality, and disperse nonnative pest species of plants and animals. At larger scales, roads affect wildlife migration patterns. In some cases, roads can also enhance roadside habitats for native species.

The importance of integrating environmental considerations into all phases of transportation is emphasized in legislation. The Transportation Equity Act for the Twenty-First Century (TEA-21) of 1998 called for protection of the environment by initiating transportation projects that would improve environmental quality and support fuel efficiency, cleaner fuels, and alternative transportation. The act called for streamlining procedures to reduce red tape and paperwork in transportation project reviews without compromising environmental protection.

Consideration of environmental issues in road development has been an evolving process. The increasing awareness of environmental issues, regulatory changes, and new solutions have made road development more complex and controversial. Many believe that environmental protection can be compatible with streamlining the project approval process through effective planning and coordination. Suggestions on how environmental and transportation goals can be better integrated have been developed by government agencies and nongovernmental organizations. Approaches include more integrated planning and interagency coordination, consideration of alternative designs earlier in the planning process, and consideration of mitigation strategies, such as installation of wildlife crossings and native vegetation management. As the road system expands and construction and management require additional resources, more is understood about the impact of roads on the environment, but much remains to be learned. To address these matters, better understanding of road ecology and improved methods of integrating that understanding into all aspects of road development are needed.

Over the past two decades, the Federal Highway Administration and state transportation agencies have increasingly recognized the importance of the effects of transportation facilities on the natural environment. The importance of this issue was reflected by congressional action in Section 5107(b)(4) of TEA-21, which required the secretary of transportation to "study the relationship between highway density and ecosystem integrity, including the impacts of highway density on habitat integrity and overall ecosystem health, and to develop a rapid assessment methodology for use by transportation and regulatory agencies in determining the relationship between highway density and ecosystem integrity." Section 5107(d) of TEA-21 authorized the secretary to arrange for a study of this relationship by the National Research Council (NRC). In response, at the request of the Federal Highway Administration, the NRC established the Committee on Ecological Impacts of Road Density (see Statement of Task in Box S-1). This committee's report attempts to provide guidance on ways to reconcile the different goals of road development and environmental conservation.

The term "road density" is frequently used to mean the average total road length per unit area of landscape. However, roads also have widely varying widths; therefore, lane miles per square mile (or lane length per unit area) is a better measure of density because it takes into account the differences between, for example, multilane expressways and two-lane rural roads. The concept of road density was developed as a way of quantifying one aspect of a road network and is applicable at scales larger than a road segment. Road density may be appropriate for measuring the structure of some existing road networks (especially those few urban or rural systems in a rectilinear grid), but it is not the only measurable term that can be used to describe road pattern and structure.

**BOX S-1** Statement of Task

A multidisciplinary committee will be established to review the scientific information on the ecological effects of road density, including the impacts of roads and highway density on ecosystem structure and functioning and on the provision of ecosystem goods and services. The committee will focus on hard-surfaced roads and will assess data and ecological indicators needed to measure those impacts. Cumulative effects will be considered. The proposed study will also provide a conceptual framework and approach for the development of a rapid assessment methodology that transportation and regulatory agencies can use to assess and measure ecological impacts of road density. To the degree that the committee can identify documentation of their effectiveness, it will consider the potential ameliorating effects of measures that might avoid, reduce, or compensate for the effects of highways and highway density on the structure and processes of ecosystems.

The committee will consider such questions as the following:

1. What are appropriate spatial scales for different ecological processes that might be affected by roads?

2. The importance of various ecological models and their appropriateness to the analysis.

3. The applicability of various ecological indicators, such as those recently recommended by the National Research Council.

4. The degree to which the national, regional, and local environmental concerns expressed in such laws as the Endangered Species Act and the Clean Water Act are relevant to the ecological effects of roads.

The study will focus on all classes of hard-surfaced roads. The committee will consider and describe as possible the various attributes of roads that have ecological significance, such as how the right-of-way is managed, surface composition, and the presence or absence of structures such as overpasses and underpasses. It will consider the importance of the pattern of road layout on ecological systems. It will not address global or regional climate effects, since they are being studied under other initiatives. However, local climate effects are appropriate in the scale of individual project design, construction, and use, and are directly related to ecosystem performance in both long- and short-term contexts.

There are cases in which the meaning of the term "road density" is clear, but often it may be difficult to make useful comparisons between the ecological effects of different types of road networks. For example, several two-lane roads that have little traffic versus fewer, eight-lane roads that are heavily traveled. Therefore, the committee focused on variables that contribute to density, such as highway length and portion of land covered, rather than strictly on density, and used the broader concept of "scale" for evaluating environmental effects.

The committee focused on the ecological effects of federally funded paved highways in urban and rural locations. The committee did not focus on urban street networks, and no consideration was given to the ecological effects of unpaved roads, such as those found in federal forests, wilderness areas, wetlands, parks, and farms, or the ecological effects of state and local roads. The committee did not address global or regional climate effects, such as how potential climate changes might affect the interactions of organisms and the environment associated with roads and vehicles or how roads and traffic might influence climate. However, local climate interactions with road ecology are considered in this report.

Developing policy choices to balance mobility, economic growth, and environmental protection goals has been important and challenging for more than 50 years. Although the committee was not charged to evaluate such policy choices, it identified the ecological effects of roads that can be evaluated in the planning, design, construction, and maintenance of roads. The committee did not address human ecological factors; nonecological factors, such as safety; efficient movement of vehicles; or protection of farmlands, publicly owned recreation lands, and scenic, historic, and cultural areas. The committee also did not address such factors as urban sprawl or suburban growth; project costs; statewide, regional, and local planning goals; and the economic viability of the communities of users.

## ECOLOGICAL EFFECTS OF ROADS

Perhaps the most noticeable ecological effect of roads is direct, vehicle-related mortality (animals killed by collisions with vehicles). Although it is not the most threatening effect of roads for most species, mortality can reduce wildlife-population densities and ultimately affect the survival probability of local populations, including endangered or threatened species, such as the Florida panther and grizzly bear. In addition to vehicle-related mortality, roads—acting as barriers to wildlife movement—may affect wildlife-population structure by disrupting breeding patterns or impairing reproductive success because they can fragment and isolate populations. In extreme cases, the resulting limitation of gene flow could result in local extirpation of a species. Properly designed mitigation measures, such as wildlife-crossing structures, can facilitate wildlife movement across roads and reconnect isolated populations. Fish movement can also be blocked by road-crossing structures, such as culverts (usually a large pipe under a road where it crosses a stream) that are improperly designed or not present at all. Some fishes avoid moving through culverts, possibly because of the increased speed of the water flow, even if there are better habitat conditions on the opposite side. Reluctance to move, for example, downstream, could contribute to isolating upstream populations and, in some cases, localized extirpations.

In evaluating the ecological effects of roads, it is important to consider the physical, socioeconomic, and legal context, as well as the ecological context. Each has spatial and temporal dimensions. The term "road-effect zone" means the distance from a stretch of road or road segment that ecological effects can be detected. The road-effect zone is usually asymmetric extending outward on either side of the road, with varying zone boundaries. The effect of distance varies, depending on the species, location, and disturbance type. For example, animals avoid roads by a distance that increases with increasing traffic volume, and that distance varies by species. Noise from high-traffic-volume roads reduces the breeding densities and distribution of many bird species within a 40- to 1,500-m zone. Increased traffic and road density negatively affect aquatic habitats and the species that depend on them. For example, wetland species diversity is negatively correlated with paved roads up to 2 km away. Other disturbances, such as heavy metals and chemical pollution, can degrade habitat quality in the road-effect zone up to 100 m and 200 m, respectively. Vehicle-generated pollutants (such as nitrogen oxides, petroleum, lead, copper, chromium, zinc, and nickel) are the primary pollutants associated with road use. Along with pollutants from spills, litter, and adjacent land uses, they accumulate on impervious roads and enter waterways via surface runoff or atmospheric deposition. Runoff contaminated with road salt can damage vegetation and potentially cause a shift in plant community structure when salt-sensitive plant species are replaced by less-sensitive species, such as cattails and common reed grass. Salt-related vegetation changes can also affect wildlife by adversely altering habitat, inhibiting road crossing by amphibian species, and causing behavioral and toxicological impacts on birds and mammals.

Similarly, air pollution from vehicle exhaust (volatile organic compounds, nitrogen oxides, carbon monoxide, and particulate matter) can alter the composition of roadside vegetation, promoting a few dominant plant species at the expense of more sensitive species, such as ferns, mosses, and lichens. This effect can extend up to 200 m from multilane highways and up to 35 m from two-lane highways.

The underlying topography, aspect (the direction a site faces or its exposure), geology, soils, ecological conditions, and land cover all influence how a road affects the environment. For example, the environmental effects of a road that does not cross a river are different from the effects of one that crosses a river several times in a few kilometers. New patterns of water runoff can develop as the local topography is altered. Aspect can influence how quickly snow and ice melt off the road and adjacent surfaces. Original topography, geology, and soils often dictate the road path and provide construction constraints or opportunities. The environmental effects of a road also depend on the prevailing land cover and use, such as wildlands, wetlands, agricultural lands, or a river valley versus a ridge. In fire-prone landscapes, a road can serve as a firebreak if the road is wide enough or as a source of fire initiation if access to the surrounding environment is increased.

Ecological productivity is influenced by roads. The roadside between the paved road and prevailing land cover often has lower productivity and different composition than the surrounding landscape (especially for roads through forests). The native habitat conditions of a roadside are frequently altered, but when the surrounding landscape is greatly altered by development, roadsides can include some of the last remaining habitats, especially for certain native plant species and some insects, birds, and small mammals. Roadside areas can also facilitate the establishment of nonnative plants transported by vehicles, among other mechanisms, including the clearing of land during road construction. Biodiversity along roads typically is different from that in the surrounding landscape. Plants along roads must survive vehicular pollution, exposure to bright sunlight, dry soils, and regular mowing. Roadside plantings in the United States once consisted of grasses and herbs (often of European origin) known to thrive in stressful conditions. Now there is an effort to plant vegetation along many highways, some of which is selected because it is native to the United States (but not always from the local area). The linear pathways of continuous terrestrial or aquatic habitat adjacent to roads can serve as corridors for animal movement. Some animals are attracted to roadside vegetation, road kill (an animal that has been killed on a road by a motor vehicle), or the light and heat often associated with roads, and other animals are deterred by disturbances in the road-effect zone.

The ecological effects of building a road typically exhibit several time lags. Some effects of road construction are not realized until several months or even decades after a road is completed as nearby trees and other plants slowly die, although the most severe (condensed and sudden) effects typically occur when construction begins. Vegetation reestablishment efforts may result in a quick pulse of plant growth after seeding and fertilization, but the new equilibrium of vegetation along roadsides usually takes some time to establish, particularly in locations with steep slopes, rocky or nonorganic substrate, or other conditions that encourage roadside erosion.

Although most of the current and foreseeable transportation projects in the United States are along established roads, the increase in traffic volume on these roads and the selection of sites for new roads bring to the forefront the potential for new ecological impacts—and associated, often delayed responses of the environment.

## Understanding and Assessing Road Effects

As described above, a great deal is known about the ecological effects of roads, even though there is need for more and better information about cumulative, long-term, and large-scale effects. The available information, much of it reviewed, summarized, and synthesized in this report, should be used in all stages of road building and maintenance, including planning.

From planning through construction stages, ecological indicators are important in assessing road effects; however, determining the broader and cumulative effects of roads and their corridors also is important and often not captured by indicators. Ecological indicators are generally developed to quantify ecological responses to a variety of factors. Several indicators have been proposed to measure or monitor ecological effects, and some of them are applicable to the effects of roads.

Ecological effects of roads at local scales (within a few kilometers of the roads) have been widely studied, documented, and understood, while effects at large scales are less documented and understood. More is known about the effects of bridges, overpasses, and culverts on flows of materials and organisms than about the effects of roads on larger patterns and processes, such as watersheds or migratory pathways. The lack of information at large scales is related to many factors, such as (1) legal and policy directives that guide what components of ecosystems must be considered; (2) planning and assessment practices that restrict scales; (3) limitations of data, indicators, and methods at broad scales; and (4) limited financial and technical support for ecological investigations at large scales.

## CONCLUSIONS AND RECOMMENDATIONS

**CONCLUSION:** Most road projects today involve modifications to existing roadways, and the planning, operation, and maintenance of such projects often are opportunities for improving ecological conditions. A growing body of information describes such practices for improving aquatic and terrestrial habitats.

**Recommendation:** *The many opportunities that arise for mitigating or reducing adverse environmental impacts in modifications and repairs to existing roads should not be overlooked. Environmental considerations should be included when plans are made to repair or modify existing roads, as well as when plans are made to build new roads.*

**CONCLUSION:** Planning boundaries for roads and assessing associated environmental effects are often based on socioeconomic considerations, resulting in a mismatch between planning scales and spatial scales at which ecological systems operate. In part, this mismatch results because there are few legal incentives or disincentives to consider environmental effects beyond political jurisdictions, and thus decision making remains primarily local. The ecological effects of roads are typically much larger than the road itself, and they often extend beyond regional planning domains.

Scientific literature on ecological effects of roads generally addresses local-to-intermediate scales, and many of those effects are well documented. However, there are few integrative or large-scale studies. Sometimes the appropriate spatial scale for ecological research is not known in advance, and in that case, some ecological effects of roads may go undetected if an inappropriate scale is chosen. Few studies have addressed the complex nature of the ecological effects of roads, and the studies that have done so were often based on small sampling periods and insufficient sampling of the range of variability in ecological systems.

**Recommendation:** *Research on the ecological effects of roads should be multiscale and designed with reference to ecological conditions and appropriate levels of organization (such as genetics, species and populations, communities, and ecological systems.)*

**Recommendation:** *Additional research is needed on the long-term and large-scale ecological effects of roads (such as watersheds, ecoregions, and species' ranges). Research should focus on increasing the understanding of cross-scale interactions.*

**Recommendation:** *More opportunities should be created to integrate research on road ecology into long-term ecological studies by using long-term ecological research sites and considering the need for new ones.*

**Recommendation:** *Ecological assessments for transportation projects should be conducted at different time scales to address impacts on key ecological system processes and structures. A broader set of robust ecological indicators should be developed to evaluate long-term and broad-scale changes in ecological conditions.*

**CONCLUSION:** The assessment of the cumulative impacts of road construction and use is seldom adequate. Although many laws, regulations, and policies require some consideration of ecological effects of transportation activities, such as road construction, the legal structure leaves substantial gaps in the requirements. Impacts on certain resources are typically authorized through permits. Permitting programs usually consider only direct impacts of road construction and use on a protected resource, even though indirect or cumulative effects can be substantial (for example, effects on food web components). The incremental effects of many impacts over time could be significant to such resources as wetlands or wildlife.

**Recommendation:** *More attention should be devoted to predicting, planning, monitoring, and assessing the cumulative impacts of* roads. In some cases, the appropriate spatial scale for the assessment will cross state boundaries, and especially in those cases, collaboration and cooperation among state agencies would be helpful.

**CONCLUSION:** The methods and data used for environmental assessment are insufficient to meet the objectives of rapid assessment, and there are no national standards for data collection. However, tools for in situ monitoring, remotely sensed monitoring, data compilation, analysis, and modeling are continually being improved, and because of advances in computer technology, practitioners have quick access to the tools. The new and improved tools now allow for substantial improvements in environmental assessment.

**Recommendation:** *Improvements are needed in assessment methods and data, including spatially explicit models. A checklist addressing potential impacts should be adapted that can be used for rapid assessment. Such a checklist would focus attention on places and issues of greatest concern. A national effort is needed to develop standards for data collection. A set of rapid screening and assessment methods for environmental impacts of transportation and a national ecological database based on the geographic information system (GIS) and supported by multiple agencies should be developed and maintained for ecological effects assessment and ecological system management across all local, state, and national transportation, regulatory, and resource agencies. Standard GIS data on road networks (for example, TIGER) could be interfaced with data models (for example, UNETRANS) to further advance the assessment of ecological impacts of roads.*

**Recommendation:** *The committee recommends a new conceptual framework for improving integration of ecological considerations into transportation planning. A key element of this framework is the integration of ecological goals and performance indicators with transportation goals and performance indicators.*

**Recommendation:** *Improved models and modeling approaches should be developed not only to predict how roads will affect environmental conditions but also to improve communication in the technical community, to resolve alternative hypotheses, to highlight and evaluate data and environmental monitoring, and to provide guidance for future environmental management.*

**CONCLUSION:** With the exception of certain legally specified ecological resources, such as endangered or threatened species and protected wetlands, there is no social or scientific consensus on which ecological resources affected by roads should be given priority attention. In addition, current planning assessments that focus on transportation needs rarely integrate other land-management objectives in their assessments.

**Recommendation:** *A process should be established to identify and evaluate ecological assets that warrant greater protection. This process would require consideration not only of the scientific questions but also of the socioeconomic issues. The Federal Highway Administration should consider amending its technical guidance, policies, and regulations based on the results of such studies.*

**CONCLUSION:** The state transportation project system offers the opportunity to consider ecological concerns at early planning stages. However, planning at spatial and temporal scales larger than those currently considered, generally does not address ecological concerns until later in a project's development.

**Recommendation:** *Environmental concerns should be integrated into transportation planning early in the planning process, and larger spatial scales and longer time horizons should be considered. Adding these elements would help to streamline the planning process. Metropolitan planning organizations and state departments of transportation should conduct first-level screenings for potential environmental effects before the development of a transportation improvement plan. Transportation planners should consider resource-management plans and other agencies' (such as the U.S. Corps of Engineers, U.S. Environmental Protection Agency, U.S. Fish and Wildlife Service, and National Park Service) environmental plans and policies as part of the planning process. Other agencies should incorporate transportation forecasting into resource planning.*

**CONCLUSION:** Elements of the transportation system, including the types of vehicles and their fuels, will continue to evolve. Changes in traffic volume and road capacity, mostly through widening of roads rather than construction of new corridors, have smaller but nevertheless important ecological effects compared with the creation of new, paved roads.

**Recommendation:** *Monitoring systems should be developed for the evaluation and assessment of environmental effects resulting from changes in the road system—for example, traffic volume, vehicle mix, structure modifications, and network adjustments. Data from monitoring could then be used to evaluate previous assessments and, over the long term, improve understanding of ecological impacts.*

**CONCLUSION:** Much useful information from research on the ecological effects of roads is not widely available because it is not in the peer-reviewed literature. For example, studies documenting the effects of roads on stream sedimentation have been reported in documents of state departments of transportation, the U.S. Army Corps of Engineers, and the World Bank. Although much of this literature is available through bibliographic databases, it is not included in scientific abstracting services and may not be accessible to a broader research community. Also, the data needed to evaluate regulatory programs are not easily accessible or amenable to synthesis. The data are typically contained in project-specific environmental impact statements, environmental assessments, records of decision, or permits (for example, wetlands permits), which are not easily available to the scientific community.

**Recommendation:** *Studies on ecological effects of roads should be made more accessible through scientific abstracting services or through publication in peer-reviewed venues. The Federal Highway Administration, in partnership with state and federal resource-management agencies, should develop environmental information and decision-support systems to make ecological information available in searchable databases.*

**CONCLUSION:** Transportation agencies have been attempting to fill an institutional gap in ecological protection created by the multiple social and environmental issues that must be addressed at all phases of road development. The gaps often occur when problems arise that are not covered by agency mandates or when agencies need to interact with other organizations in new ways. Even when transportation agencies work toward environmental stewardship, they cannot always do the job alone.

**Recommendation:** *Transportation agencies should continue to expand beyond their historical roles as planners and engineers, increasing their roles as environmental coordinators and stewards. Transporta-*

*tion planners and natural-resource planners should collaborate to promote integrated planning at comparable scope and scale so that the efforts can support mutual objectives. This collaboration should include federal, state, and county resource-management agencies; nongovernmental organizations; and organizations and firms involved in road construction. Incentives, such as funding and technical support, should be provided to help planning agencies, resource agencies, nongovernmental groups, and the public to understand ecological structure and functioning across jurisdictions and to interact cooperatively.*

# 1

# Introduction

For at least the past four millennia, human development has been intertwined with the development of roads. Two millennia ago, the Roman culture made dramatic leaps in technological innovation to facilitate transportation and expand its empire. The Romans built straight, narrow roads to provide a stable base for moving humans, animals, and vehicles. Their roads extended long distances, creating one of the first networks of roads. With the development of roads and road networks, a variety of interactions and ecological effects occurred. Their techniques of excavation (including the removal of mountainsides), construction of bridges, and changes in water flows affected the environment and wildlife (and human) mortality. Many of these effects persist today.

The past century has seen a transformation in the magnitude and the scale of paved roads in the world in general and in the United States in particular. This transformation parallels the production and use of motorized vehicles. Roughly a century ago, the first fossil-fueled vehicles were being driven on a road system where the patterns on the landscape were largely in place but undeveloped. In 1900, an extensive network of roads existed, but only about 4% were paved (FWHA 1979). In the twentieth century, the system about doubled in length (Forman et al. 2003) to a recent estimate of 4 million linear miles or about 0.8% of the land surface area in the United States (Cook and Daggett 1995). As the system was getting longer, it was also becoming more structurally complex to provide for larger volumes of traffic and heavier loads. Those complexities include larger, wider roads with changing techniques of construction and greater structural stability. Some of these changes were done, in part, to accommodate and mitigate environmental issues.

As the road system became more complex in the twentieth century, it became more expensive to build, repair, and maintain. Road development in the early 1900s relied on local or private sources of funding. Funding is now derived from a variety of sources, and a large percentage originates from federally collected sources, particularly user-related fees on gasoline and transport equipment. Since the passage of the Transportation Equity Act for the Twenty-First Century (TEA-21), investment in highway infrastructure has increased; for example, investment increased by 14% between 1997 and 2000 (FHWA 2002a). Capital spending on highways totaled about $64 billion in 2000, but that amount is unlikely to meet future needs for maintenance (FHWA 2002a).

The process by which these monies are spent to administer, plan, select, and build projects and maintain roads is a complicated maze of bureaucracy. The process is centered in states—with some variation among states in planning and implementation—and involves five interacting groups: federal government, state government, metropolitan planning organizations, local government, and the public (Forman et al. 2003). These governments share funding of administration, construction, planning, and maintenance in various ratios. Governments construct and maintain roads to provide a number of social benefits, such as providing mobility, transporting goods and people, and sustaining economic growth. One of the considerations in the process of road development is the recognition and management of environmental concerns.

All phases of road development—from construction to vehicle use—change ecological conditions of an area. Roads change abiotic characteristics of the environment—from physical to chemical soil conditions to alterations in water flow and water quality. Vehicle use on roads contributes to air and water quality degradation (Forman et al. 2003). Other direct changes to the biotic components of ecosystems include alteration of habitats, increased wildlife mortality, and dispersal of nonnative pest species (plants and animals). At larger scales, roads can affect migration patterns (Forman et al. 2003).

Integration of environmental effects into all phases of transportation efforts has been part of an ongoing process for several decades. These efforts attempt to form a bridge between the transportation policy arena and the environmental science community. Federal legislation has long directed the Federal Highway Administration (FHWA) and recipients of federal funding to consider environmental factors when planning transportation systems and specific road projects. During the 1990s, the Intermodal Surface Transportation Equity Act and the TEA-21 (FHWA 2001a) contained provisions addressing environmental considerations for both transportation planning and project implementation. Combined with programs to address cleaner fuels and cleaner vehicle emissions, many programs have aimed at addressing and reducing the environmental impacts associated with roads and the mobility roads provide. At the same time, there have been efforts to streamline environmental review processes to reduce red tape, paper work, and delay in project reviews without compromising environmental protections. Striking a balance between competing factors and policies has been and remains a daunting challenge.

The consideration of environmental issues in the road development process has been a source of debate. Some argue that dealing with environmental concerns adds unnecessary time and cost to road projects (GAO 2003). Others contend that it is just one of many factors that protract projects. Recent work by FHWA suggests that streamlining environmental considerations is possible through effective planning and coordination.

Research on road effects has been conducted for decades, but a larger emphasis has been placed on investigating road ecology in the past decade. Forman et al. (2003) provided a compilation of the ecological effects of roads, assessing the direct and indirect effects on vegetation, animals, and abiotic influences across a range of scales. They also argued for the development of a discipline of road ecology that combines an improved understanding of ecological effects with all dimensions (assessment, planning, coordination, construction, and management) of road development.

Suggestions on how environmental and transportation concerns can become more integrated have also been addressed by governmental and nongovernmental agencies. In a report published by the Transportation Research Board, Evink (2002) suggested that many issues associated with wildlife can be addressed through early planning or context-sensitive design, where structural solutions can mitigate some effects. White and Ernst (2003) recommended a number of ways to mesh conservation and transportation objectives, including integrated planning, conservation banking, agency coordination, wildlife crossings, and native vegetation management.

As more resources go toward road construction and management and as knowledge about the ecological effects of roads increases, the conflicts between the societal goals of developing transportation infrastructure and maintaining ecological goods and services (for example, soil production [a good] from decomposition of plant matter [a service]) become more apparent. Reconciling the conflicts is made more difficult by the lack of knowledge of some ecological changes caused by roads—especially local, site-specific changes and processes at all appropriate ecological scales—and by the demand for rapid, efficient use of human resources for road development. The question is how can efficient use of resources in road planning, construction, and use be increased while attempting to conserve critical ecosystem structure, functioning, and services?

## COMMITTEE CHARGE AND RESPONSE

The increasing importance of the relationships of transportation facilities to the surrounding natural communities and their wildlife has been recognized by FHWA and state transportation agencies. The importance of transportation corridors and infrastructure was pointed out by the National Research Council (NRC 2000a) when it called them globally significant. The importance of this issue was reflected by congressional action in Section 5107(b)(4) of the TEA-21 (PL 105-178), which required the secretary of transportation to "study the relationship between highway density and ecosystem integrity, including the impacts of highway density on habitat integrity and overall ecosystem health, and develop a rapid assessment methodology for use by transportation and regulatory agencies in determining the relationship between highway density and ecosystem integrity." Section 5107(d) of TEA-21 authorizes the secretary to fund a study of this relationship by the National Academies through a grant or cooperative agreement. In response, FHWA asked the National Academies to direct its investigative arm, the NRC, to review the scientific information on the ecological effects of road density. See Box 1-1 for details of the charge to the committee and Box 1-2 for the committee's definition of paved roads.

In 2002, the NRC formed the Committee on Ecological Impacts of Road Density, a panel of 14 members that included a experts in environmental engineering, highway construction and engineering, land-use change, wildlife ecology, endangered species, habitat evaluation modeling, habitat impact assessment, economic development and planning, environmental law and policy, and biodiversity conservation (see Appendix A).

The committee held three public meetings—Washington, DC, Irvine, CA, and Seattle, WA—to collect and review available scientific information, hear briefings from scientists and transportation profession-

---

**BOX 1-1** Statement of Task

A multidisciplinary committee will be established to review the scientific information on the ecological effects of road density, including the impacts of roads and highway density on ecosystem structure and function and on the provision of ecosystem goods and services. The committee will focus on hard-surfaced roads and will assess data and ecological indicators needed to measure those impacts. Cumulative effects will be considered. The proposed study will also provide a conceptual framework and approach for the development of a rapid assessment methodology that transportation and regulatory agencies can use to assess and measure ecological impacts of road density. To the degree that the committee can identify documentation of their effectiveness, it will consider the potential ameliorating effects of measures that might avoid, reduce, or compensate for the effects of highways and highway density on the structure and processes of ecosystems.

The committee will consider such questions as the following:

1. What are appropriate spatial scales for different ecological processes that might be affected by roads?

2. The importance of various ecological models and their appropriateness to the analysis (e.g., NRC 2000a).

3. The applicability of various ecological indicators, such as those recently recommended by the National Research Council (NRC 2000b).

4. The degree to which the national, regional, and local environmental concerns expressed in such laws as the Endangered Species Act and the Clean Water Act are relevant to the ecological effects of roads.

The study will focus on all classes of hard-surfaced roads. The committee will consider and describe as possible the various attributes of roads that have ecological significance, such as how the right-of-way is managed, surface composition, and the presence or absence of structures, such as overpasses and underpasses. It will consider the importance of the pattern of road layout on ecological systems. It will not address global or regional climate effects, since they are being studied under other initiatives. However, local climate effects are appropriate in the scale of individual project design, construction, and use and are directly related to ecosystem performance in both long- and short-term contexts.

---

als, and receive oral and written testimony from the public. The meetings included field site visits to transportation projects. The committee met two more times in executive session to complete its report.

---

**BOX 1-2** Definitions of Paved Roads

A paved road is any road that is described by the following categories (FHWA 2002a):

LOW TYPE: an earth, gravel, or stone roadway that has a bituminous surface course less than 1 inch (in.) thick suitable for occasional heavy loads.

INTERMEDIATE TYPE: a mixed bituminous or bituminous penetration roadway on a flexible base having a combined surface and base thickness of less than 7 in.

HIGH-TYPE FLEXIBLE: a mixed bituminous or bituminous penetration roadway on a flexible base having a combined surface and base thickness of 7 in. or more; it also includes brick, block, or combination roadways.

HIGH-TYPE COMPOSITE: a mixed bituminous or bituminous penetration roadway of more than 1 in. compacted material on a rigid base with a combined surface and base thickness of 7 in. or more.

HIGH-TYPE RIGID: a hydraulic cement concrete roadway with or without a bituminous wearing surface of less than 1 in.

---

The focus of the committee's work was on the ecological effects of federally funded, major paved roads within the United States road network—for example, highways in urban and rural locations. The committee limited its focus to highly urbanized street networks. The committee recognizes that paved roads run through a range of conditions, from undeveloped areas (such as, parklands or conservation areas) through rural and agricultural lands to high density, urban development. More is known of the ecological effects of roads in undeveloped and rural areas than in highly urbanized areas. Some effects, such as air pollution and noise, increase with development. Legal requirements, policy frameworks, planning and assessment that address ecological issues also vary by context. Hence, some topics in this report focus on undeveloped and rural areas, and others focus on urban settings.

It was outside on the committee's charge to consider the ecological effects of large networks of roads that are not paved, such as those found in federal wetlands, forestlands, wilderness areas, farm roads, or state-level categories. Global climate effects, either how potential climate changes might effect road ecology or how road ecology might influence climate were also not included in the committee's charge. Local (smaller than meso-scale) climate interactions with road ecology are included in this report.

The committee was not formed to address the diversity of political, social, and economic factors that relate road development and traffic flow to urban sprawl or suburban growth, and these issues are not addressed in the report. The committee also did not consider the extremely important topic of human ecology in this report. Human ecology includes social, cultural, economic, and political dimensions, and it is important in rural and urban environments. The disruption of human communities—especially in urban areas—is well documented, but roads can profoundly affect humans even in remote, sparsely populated areas such as Alaska's North Slope (e.g., NRC 2003). Roads obviously have enormous effects on commerce, many of which are by design, but many of their effects are unintended or indirect. The topic is broad and complex enough to warrant its own committee report. Such a committee would be constituted very differently from this one, and that is the main reason that the topic is not discussed in detail in this report. The committee was not constituted to address other important quality-of-life factors such as safety; efficient movement of vehicles; and protection of farmlands, publicly owned recreation lands or scenic, historic, and cultural areas. Nor does it address such important considerations as project and other related costs; statewide, regional, and local planning goals; and the economic viability of the communities of users.

This report strives to highlight the ecological effects of highways that should be evaluated throughout the decision-making process in the planning, design, construction, and maintenance of highways.

The term *road density* is used in the charge to the committee. Forman et al. (2003) defined the term as "the average total road length per unit area of landscape," which is intuitively appealing because it is easy to measure. That definition is used fairly widely in the literature on the ecological effects of roads. However, roads also have width, which can vary widely, and therefore lane miles per square mile (or lane length per unit area) is a better measure of density. That measure takes into account the differences between major multilane expressways and two-lane rural roads, as well as any road in between. Therefore, when this committee discusses road density in general, lane length per unit area is meant. However, it often is more difficult to obtain information on lane length than on road length, and many studies considered in this report use road length only. In specific cases where other studies are cited, the definition of road density used in those studies is specified.

The concept of road density was developed as a way of quantifying one aspect of a road network, and therefore the term is applicable only at scales larger than a road segment or for a system of roads. The term road density may be appropriate for measuring the structure of some existing road networks (especially those few urban or rural systems in a rectilinear grid), but it is not the only measurable term that could be used to describe road pattern and structure. The term may also be useful for assessing some ecological effects (such as a surrogate of impervious surface).

However, in attempting to evaluate road density, the committee came to recognize that the term does not refer to a simple or single variable. *Density* includes length, width, number of intersections, and other related variables. Although the meaning of the term was reasonably clear in some cases, such as in its relevance to the number of lanes in a highway, the committee often had difficulty identifying a good way to make comparisons. For example, the concept of road density alone does not allow a useful comparison of the effects of two road networks, one of which consists of many narrow roads that have little traffic and one of which has fewer roads that are wider and heavily traveled. In another example of using the concept of density alone, it is unclear how the effects of one 8-lane highway can be compared with those of two 4-lane highways. In the above comparisons and in many others, the roads must be observed on the ground and their specific effects studied.

Therefore, the committee has not devoted a great deal of attention in this report to density per se but has focused on variables that contribute to density, such as highway length, portion of land covered, nature of interchanges, and so on. It also has used the broader concept of scale for evaluating environmental effects. In a few cases, however, density is a tractable and useful way to think about the ecological effects of roads.

The committee was asked to consider the cumulative effects of roads. *Cumulative impact* is defined by the Council on Environmental Quality as "the impact on the environment which results from the incremental impact of the action when added to other past, present, and reasonably foreseeable future actions regardless of what agency (Federal or non-Federal) or person undertakes such other actions." Cumulative effects can result from individually minor but collectively significant actions taking place over a period of time. The term ensures that environmental impacts of federal actions, such as transportation projects, are considered in connection with other activities that can affect the same environmental resources. In this report, the committee considers the concept of cumulative effects more broadly than the regulatory definition to look at effects that arise from synergistic interactions, involving multiple factors and occurring over large spatial and long time scales. In assessing and managing cumulative effects, especially in response to legal and regulatory requirements, this committee agrees with an earlier NRC (2003) committee that it is helpful to focus "on whether effects under consideration interact or accumulate over time and space, either through repetition or combined with other effects, and under what circumstances and to what degree they might accumulate." For an extensive discussion of cumulative effects assessment, see NRC (2003).

*Ecosystem goods and services* are terms used to describe those structures or processes that provide support for a variety of human endeavors (Daily 1997). This utilitarian construct, originally described by Daily (1997), defines components of ecological systems in terms of how they benefit and support human life. Ecosystem services include water purification, flood and drought mitigation, climate stabilization, carbon sequestration, waste treatment, biodiversity conservation, soil generation, disease regulation, pollination, maintenance of air quality, and the provision of aesthetic and cultural benefits. Ecosystem goods are produced by these services and include food, fiber, timber, genetic resources, and medicines. Most ecosystem goods are direct inputs into economic systems. The committee assumed that ecosystem goods and services are fairly well correlated with ecological structure and functioning. Some of the report addresses how these goods and services interact with road ecology. For the most part, however, the report focuses on ecosystem structure and functioning.

In evaluating the ecological effects of roads, the physical, socioeconomic, and legal contexts in which the roads are situated are important. Each context has spatial and temporal dimensions.

The spatial context is largely determined by the physical conditions of the environment. Topography, aspect (a position facing in a particular direction or exposure), underlying geology, stream network, and soils all influence how a road affects the environment. For example, the interaction between an environment and a road that does not cross a river is different from that between an environment and a road that crosses a river six times in 10 kilometers (km). Aspect can influence how quickly snow and ice melt off the road surface and thereby affect driving conditions, which influence both traffic speed and volume. Rapid snowmelt could also affect stream hydrographs. Original topography, geology, and soils dictate the road path and provide construction constraints or opportunities.

Roads change the local physical environment by interacting with underlying topography, aspect, geology, soils, ecological conditions, and land cover. For example, new patterns of water runoff can develop as the local topography is altered, and those changed patterns can result in altered storm hydrographs, changed groundwater recharge, and increased delivery of sediments and contaminants.

The effects of a road also depend on the prevailing type and intensity of land cover and land use. Environmental effects are largely influenced by whether a road runs through wildlands (removing natural habitats and opening up new areas to human access), wetlands, agricultural lands, or a river valley or whether it lies on a ridge. In fire-prone landscapes, roads can serve as a firebreak if the road width is enough to deter spread of ground fires.

Ecological productivity is also influenced by the presence of roads. The roadside between the paved road and original land cover often has lower productivity than the surrounding landscape (especially for roads through forests). Roadside maintenance is designed for driver safety and often involves the use of chemical deterrents or mechanical removal of vegetation. The linear pathway along a road can serve as a corridor for movement of native and nonnative plants and animals. For example, some highways in Georgia are bordered by the kudzu vine, which was introduced and grows into the native pine canopy and eventually kills the trees.

Biodiversity along roads typically is quite different from the surrounding landscape. Plants along roads tolerate vehicular pollution, exposure to bright sunlight, dry soils, and regular mowing. Roadside plantings in the United States once consisted of forbs and herbs (often of European origin) known to thrive in extreme conditions. Now there is an effort to plant vegetation along many highways, some of which are selected because they are native to the United States (but not always from the local area). For example, California poppies are now abundantly grown outside their original range. Sometimes roadsides function to protect native plant communities and may be the only way some plants are protected from land alteration. Some animals use habitats found near roads, such as deer feeding on vegetation, snakes basking, or many animals feeding on road kill (an animal that has been killed on a road by a motor vehicle). Some animals appear indifferent to a road, and other animals shy away from the noise and chemical pollution from surfaces and vehicles.

To capture the diverse effects of roads on their environment, Forman et al. (2003) refer to a "road-effect zone" over which significant ecological effects occur. Those effects can include changes in the abundance (or even presence) of plants and animals; barriers to movement of both terrestrial and aquatic animals; changes in water levels, flows, and water quality; and other habitat changes that affect populations and biological communities. Because these factors vary over space, the road-effect zone is usually asymmetric extending outward on either side of the road, with varying zone boundaries.

Temporal changes on the land are influenced by roads. The ecological impact of building a road occurs with several time lags in response. Some effects of road construction are not realized for several decades after a road is completed, when trees and other plants die and wildlife mortality affects population persistence. Reestablishment efforts may result in a quick pulse of plant growth after seeding and fertilization, but the new equilibrium of vegetation along roadsides may take some time to establish.

The existing road system and the addition of new roads to that system often have different impacts and management options. Although most current and foreseeable road projects in the United States are along established roads, the selection of sites for new roads carries the potential for new ecological effects. New roads can affect the movement of plants and animals and change the physical environment, as can increases in traffic volume on existing roads.

Ecological indicators are important in assessing the effects of roads during planning through construction stages of road projects, and they are important in determining the broader and cumulative effects of the road and its corridor. Ecological indicators are generally developed to quantify the ecological response to a variety of factors (Hunsaker and Carpenter 1990, Suter 1993) and are further discussed in Chapter 7.

## WHAT IS DIFFERENT AND NEW IN THIS REPORT?

This report attempts to provide guidance on reconciling the different goals of road development and of environmental conservation in two ways. The use of scale is an important factor in understanding the ecological context of roads, and the integration of social and ecological dimensions is important in assessing, constructing, and managing road ecology. These themes are discussed in the following sections.

## Scales of Observation

In traditional use, the word *scale* has at least two meanings. One defines a unit of measurement. A meter and a foot are different scales, and measures of objects are made using multiples or fractions of these units. For example, the paved surface of a typical two-lane highway is about 26 feet (ft) or 8 meters (m) wide. The other definition of scale has to do with relationships among units and is derived from the Latin word *scalaris* for ladder, which is used for such items as the scale of a map. For example, the scale of a road map is 1:50,000, where 1 inch (in.) on the map equals 50,000 in. on the ground.

In this report, two terms, *grain* and *extent* (O'Neill et al. 1986), are associated with scale. A grain is defined as the unit of the smallest resolution within a data set and is similar in meaning to the first definition of scale in the preceding paragraph. Grain and extent can be applied along either temporal or spatial dimensions and thus are useful descriptors in assessment and planning. In one-dimensional spatial data, such as a transect, the grain is the unit of measure or step length. In two-dimensional spatial data, such as a map, a pixel or grid cell size is the grain of the data set.

The extent of a data set defines the bounds in space or time. The extent could be the length of a transect for one-dimensional spatial data. For two-dimensional spatial data, the extent is defined as the window of the map. In temporal data, the grain is usually defined as the minimal time unit, such as minute, day, or year, and the extent is the period of record used in analysis. Therefore, scale is defined here by two components: the grain and extent.

As a demonstration of these meanings of scale and its relevance to road ecology, consider the set of Figures 1-1 through 1-8. The concept of viewing the world at different scales was originated by Boeke (1957) and can be found in a book and video called *Powers of Ten* by Morrison and Morrison (1982). Both works generated a set of images that were differentiated in size by an exponent of 10 ($10^1$, $10^2$, $10^3$, and so forth). In this sequence, there is a photograph of the center of a road on the campus of Emory University in Atlanta, Georgia (Figure 1-1). The extent of the photograph is 1 m, a grain being about 2 millimeters (mm).

Each subsequent image increases the extent by an order of magnitude or power of 10 (for example, 10 m, 100 m, 1 km, and so forth) while retaining roughly the same central point. The sequence ends with an image of North America—an extent of about 10,000 kilometers (km). The set of images provides a glimpse into the complexity of the ecosystems at each of the scales.

Figures 1-1 through 1-8 depict distinct changes in structures as the scale of observation changes over 8 orders of magnitude. At the smallest or finest scales, the road surface is visible (Figure 1-1). Road structures—markings, sidewalks, curbs—are apparent in the next scale (Figure 1-2). As the scales become larger, buildings, parking garages, athletic fields appear (Figure 1-3); then road networks appear through campus and suburban neighborhoods and metropolitan features appear—areas of intensive development that are linked by roads (Figure 1-4); road patterns are evident, yet individual roads are scarcely visible (Figure 1-5); land uses of urban areas, suburban housing, agricultural fields, and forests appear (Figure 1-6); geological and hydrological features, such as mountains ranges and coastlines, with large urban areas are still visible (Figure 1-7); and finally, geomorphological structures and land, ocean, and atmosphere interactions that mediate climate change and sea-level change appear (Figure 1-8).

Examination from the perspective of a square meter in the middle of a road to that of a continent reveals three observations. First, as the extent and grain of scale increases, distinct objects appear and persist over distinct scale ranges and are replaced by others that are either aggregates of objects at smaller scales or new objects. For example, the double strip marking in the center of the road is no longer visible at extents of a kilometer, and road segments visible at smaller extents aggregate to become road networks in larger extents. At each such range of scales, the identifiable objects have geometric properties. For example, the metric of road density is only applicable at scales with extents greater than about 1 km. The second observation is that the patterns and structures change across scales; that is, these complex systems are not self-similar and amenable to simple scaling relationships. Ecological structures persist over given scale ranges, then change as the sets of processes that organize those structures change.

For example, the effects of runoff from a road on accumulation of toxic material and subsequently on individual plants may include several scale ranges.

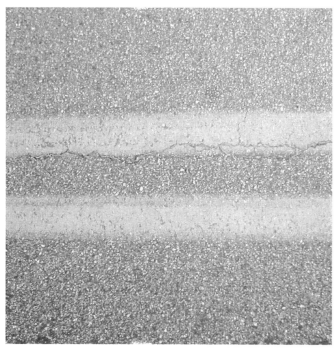

**FIGURE 1-1** Photograph of double yellow line in the center of a road, Emory University campus, Atlanta Georgia. Window size (extent) of photograph is 1 m × 1 m. Source: Photograph by L. Gunderson.

The chemistry of interactions among compounds occurs at small scales (orders of magnitude smaller than the first image (Figure 1-1). Effects of toxins on individual biota are evaluated at scales depicted in Figure 1-1 or 1-2. At scales of the region (Figure 1-7), local accumulations of toxins may be undetectable. The third observation is that the effects change with scale and medium (water, air, or land). Human effects on the atmosphere cross a wide range of scales—exhausts from individual vehicles can accumulate and aggregate to have planetary effects on atmospheric carbon dioxide and other greenhouse gases. Human influence on land, however, appears more local. For example, roads and even road networks affect hydrological and geological processes but only at scales up to a watershed and not at continental scales.

The sequence of photographs suggests that road structures and attributes vary with scale. The attributes of paved roads (materials of asphalt, gravel, and concrete) and thicknesses are described in spatial extents of observation from millimeters to centimeters. At scales of centi-

**FIGURE 1-2** A larger segment of the road as seen in Figure 1-1. The entire width of the road, sidewalks , and shadows of trees are visible. Extent of photograph is 10 m × 10 m. Source: Photograph by L. Gunderson.

meters to meters, road markings, curbs, drains, and culverts are objects of design and construction. At scales of a few meters, the structures of pavement width, shoulders, travel lanes, and paved envelope are germane. Many of the ecological effects are contained within a kilometer of the paved roads, an area often referred to as the road-effect zone. At scales of multiple kilometers, road densities, patterns, traffic use, and alignments are elements of the road system.

These simple models of scales—especially scales at which roads affect ecological systems—and the integration of different disciplinary approaches to road ecology lay the foundation for the discussions in this report.

### Integration of Ecological and Social Systems

Perhaps one of the most difficult challenges facing society is the integration of human-development activities and ecological-resource con-

**FIGURE 1-3** A larger (longer segment) of road on Emory campus. The road divides a parking deck on top and athletic field on bottom. Extent of photograph is 100 m × 100 m. Source: USGS 2004.

servation. This challenge is particularly true of all phases of roads—from construction to use to removal—as all of these actions interact with and alter ecological systems (Forman et al. 2003).

Ecological systems are defined as systems comprising biotic (organisms) and abiotic (physical) components (Odum 1983). The biotic components interact with abiotic components (such as solar energy, water, air, and nutrients) in ways that generate complex and diverse structures. The interaction between structures and processes has been described as self-organization (Odum 1983; Levin 1999), where structure and process mutually reinforce one another. In the context of this report, the committee uses the term ecological systems to describe organizational units that cover a wide range of spatial scales, from centimeters (such as a Petri dish) to thousands of kilometers (the Pacific Ocean). Even though ecological systems are not restricted to any particular temporal or spatial scale, the spatial scale and time are two key dimensions (measurable extent) for understanding ecological systems. The components of a generic forest ecosystem also cover wide ranges of scale, as

**FIGURE 1-4** At this scale of photograph, most of the buildings that make up the central campus of Emory University can be seen embedded in a network of roads. The neighboring suburban areas border the campus. Extent of photograph is 1 km × 1 km. Source: Photograph courtesy of Emory University Facilities Management.

shown in Figure 1-9. Some of these structures cover only a distinct range of scales due to processes that limit size. Cells, leaves, and trees are limited in size. Other structures and processes have wide ranges of scale. Emissions from the combustion of fossil fuels can aggregate to produce global scale effects.

A human social system is defined as a group of people who share understanding, norms, and routines to accomplish activities or fulfill key functions (Westley 2002). They may be organized to achieve goals or objectives or fulfill other needs. Ecological systems are organized around space and time dimensions, and human systems are organized around the number of people. They may be as small as a family of two or as large as a nation.

There are several ways to conceptualize the relationship between ecological and human social systems.

The most appropriate approach for this report is a "linked" system perspective and suggests that ecological and human components have a

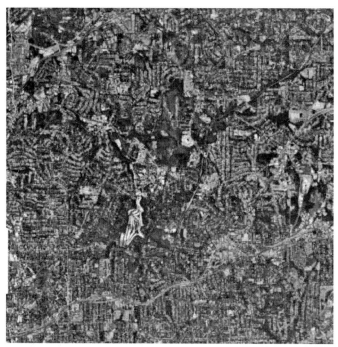

**FIGURE 1-5** At an extent of 10 km × 10 km, the road patterns around Emory campus are evident, yet individual roads are scarcely visible. The high-density suburb of Atlanta indicates a mix of land-use types and economic activity. Source: Photograph courtesy of Emory University Facilities Management.

set of rules and structures and that it is most important to focus on link-ages and feedbacks between these components. Berkes and Folke (1998), NRC (1999), and Gunderson and Holling (2002) all underscore the need for new science and approaches to understand the dynamics and complexities of these linked systems of people and nature. For example, Sutter (2002) traces the roots of the wilderness movement as a response to road construction in the early twentieth century. The committee adopts a linked perspective in this report and develops it in subsequent chapters.

## ORGANIZATION OF REPORT

This report addresses and is limited to the statement of task as agreed on by the NRC and FHWA. To address how ecological consid-

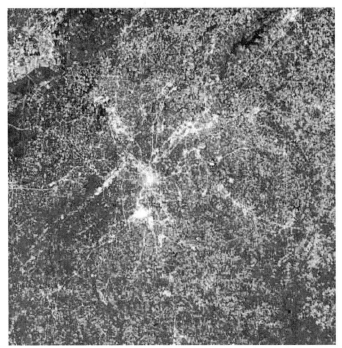

**FIGURE 1-6** At an extent of 100 km × 100 km, the urban imprint of Atlanta is visible. The light (white) areas are highly reflective areas of concrete and pavement. Traces of roads are still visible, as are the forest covers (dark) and agricultural areas. Source: USGS 2004.

erations could better be integrated into all phases of road development—assessment, planning, construction, and management—the remainder of this report is structured in seven chapters.

Chapter 2 provides a brief history and baseline description of the current state of the road system in the United States. The description includes function definitions of types of road systems (interstate, arterials, and collectors), the layout or patterns of roads, and ownership and maintenance responsibilities. Chapter 2 also describes the current status of pavement, bridge conditions, and future projections of spending and other required capital investments.

Chapter 3 addresses the effects of roads on ecological conditions by using spatial scale to sort ecological effects. It also examines how ecosystem goods and services are altered by road activities. The focus of the chapter is a literature review of the documented changes in ecological structure and functions (and goods and services) by scale of impact. A

**FIGURE 1-7** At an extent of 1,000 km × 1,000 km, geological features dominate. The large urban areas of the southeastern United States are seen as white clusters. Rivers, coastlines, and mountains are also seen. Source: USGS 2004.

large part of the scientific knowledge of ecological effects of roads has been based on short-term studies focused on narrowly defined objectives and have generally been related to specific construction or planning needs. As a result, little is known about ecological effects that occur over large spatial areas or long (decades) time frames. Ecological conditions are affected not only by the construction of the road and road appurtenances (bridges) but also by the traffic on the road and, at larger scales, by increases in road density. The committee's review has shown the ecological effects of roads to be much larger than the road itself and can extend far beyond regional planning domains. Many studies have failed to address the complex nature of the ecological effects of roads. Studies assessing ecological effects are often based on small sampling periods and therefore do not adequately sample the range of variability in ecological systems. Little is known about how roads affect the different components of biodiversity (genetic, species and population, community, and ecosystem). Information on the resiliency of biodiversity compo-

**FIGURE 1-8** At an extent of approximately 10,000 km × 10,000 km, clouds and atmospheric structures appear as well as the entire continent of North America. Source: Davis and Ogden 1994. Reprinted with permission; copyright 1994, St. Lucie Press, Delray Beach, FL.

nents to road-related disturbances is needed to better understand the effects of roads on ecological systems.

Chapter 4 outlines existing and potential opportunities at different scales for mitigating the ecological effects associated with three major phases of road projects: planning, construction, and maintenance.

Chapter 5 reviews the existing laws, regulations, and policies and their influence on the interaction between ecology and roads. Although a wide range of laws, regulations, and policies require some degree of consideration of ecological effects of road construction, the existing legal structure leaves significant gaps. Road projects need only permits to affect certain types of ecological features—wetlands, endangered species, and migratory birds—and generally at a small scale. Moreover, the permit programs generally only consider direct effects of road construction or use on the protected resource. With few exceptions (for example, wetlands and endangered species), existing law authorizes ecological con-

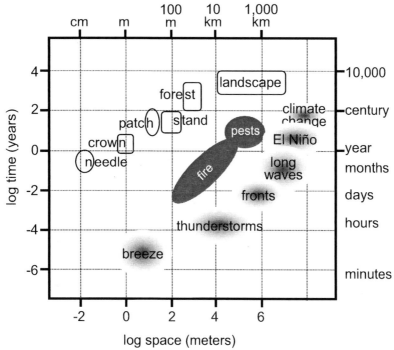

**FIGURE 1-9** Ecological hierarchy indicated by the spatial and temporal scales of vegetation structures (needles, patches, and forests), disturbance processes (fire), and atmospheric structures (fronts).

cerns to be balanced with goals of transportation mobility, capacity, and other social needs in determining whether and how to undertake transportation projects. This can create controversy between supporters of a project and opposers who raise issues of ecological protection. These are primarily federal level acts that either directly or indirectly influence activities at smaller spatial scales.

Chapter 6 addresses the current practices of planning and assessing road projects. Large-scale planning processes (such as long-range transportation plan) are required to address only air quality issues (such as attainment of standards) and generally do not address other environmental issues. Integration of environmental concerns into transportation planning should be done earlier in the planning processes. Such organizations as the metropolitan planning organizations should conduct first-level screenings for environmental considerations in transportation improvements before the development of a transportation improvement program (TIP). TIP is a fiscally balanced, itemized list of all federal and regionally important state-funded transportation projects planned for the metropolitan area usually covering 2 years. Federal transportation agencies, such as FHWA and Federal Transit Administration, require that all projects using federal funds come from an adopted transportation improvement program. Transportation and conservation planning at the state level should be integrated. Further development of a first-level screening assessment (rapid assessment) should be conducted for use early in the planning process. To streamline environmental assessments, two steps are needed: (1) more spatial and temporal environmental data should be gathered, and (2) a set of models must be developed for using those data in assessments to address concerns dealing with scale, feedbacks, and mixed criteria for environmental protection. Transportation agencies have an opportunity to be information brokers and to foster planning forums that integrate environmental planning and assessment across governmental agencies, nongovernmental organizations, stakeholders, and the public. This chapter describes the new and emerging technologies that could be used to improve the practices. It also describes some conceptual approaches to achieve better integration of social and ecological objectives and to assess environmental concerns of road projects.

Chapter 7 attempts to integrate the findings from the previous chapters, and Chapter 8 presents the conclusions and recommendations that are identical to those in the summary.

Ecological systems cover wide ranges of spatial and temporal scales. The complexity and cross-scale interactions in ecological systems generates problems for assessment and planning. Multiple assessments must be developed, each at different spatial and temporal scales to address key ecosystem processes and structures. Ecological concerns should be included early in the assessment and planning processes. Although great progress has been made in understanding and mitigating effects of roads, much more is needed. The development of a broader set of robust ecological indicators and learning-based institutions will help facilitate assessment and planning. Better integration of the social institutions will likely require the development of new relationships among existing institutions. Transportation agencies have an opportunity to play a key role in interconnecting and integrating planning and management of environmental issues. New types of institutions are needed to address the mix of socioeconomic and ecological concerns.

# 2

# History and Status of the U.S. Road System

## INTRODUCTION

The road system in the United States has evolved over time to a complex network of physical structures that include roads, bridges, and overpasses, all designed to carry an enormous amount of traffic. The system has been created and continues to be changed and maintained by an equally complex set of human systems, centered on a hierarchy of governmental agencies with their associated financial support. The road system provides unlimited access for millions of Americans. The current system of paved roads handles a volume of traffic on the order of 2.9 × $10^{12}$ vehicle miles per year,[1] or about 8 billion vehicle miles per day (DOT 2003).

The first section of this chapter presents a brief history of the system. The second section defines terms relevant to the system and their spatial patterns at different scales. The third section describes ownership and maintenance responsibilities. The fourth section assesses the current status of the road system, which is still undergoing modest growth and requires many resources to maintain, operate, and repair.

## A BRIEF HISTORY OF THE U.S. ROAD SYSTEM

We provide this brief historical overview for three reasons: (1) to show that the layout of the current road system is unlikely to change

---

[1] One vehicle traveling 1 mile constitutes a vehicle mile.

dramatically and that most development will be done along the current spatial template; (2) to show that the road system is carrying an increasingly heavy load, thus increasing congestion; and (3) to show that increased maintenance is required because of the aging road system. As we point out later, maintenance provides opportunities for mitigating or reducing the adverse ecological effects of roads, and such opportunities should be taken advantage of.

A large and extensive road system was already in place in the United States when cars became a major mode of transportation in the early twentieth century. The pattern of the system mirrored land uses and transportation corridors of the nineteenth century. Roads were narrow, primarily composed of dirt and gravel, and for the most part, followed existing topography. Before 1900, only 4% of the roads were paved, leading to poor and unreliable traveling conditions. Yet this system formed the template for the current system. Indeed, the road system has less than doubled in length since 1900, but the capacity has multiplied to accommodate an ever-increasing demand (Forman et al. 2003).

The development of the road system occurred in distinct eras, paced in part by technological transportation developments and resource availability. Each era marked a distinct change in a suite of variables (public values, policy, and fiscal resources) that influence road development.

The historical context for roads is an important consideration because history affects the current ecological effects of roads. For example, the designers of a modern interstate highway would be more likely to be sensitive to the hydrological and ecological effects of the project than the designers of a two-lane rural road built with county funds or 50 years ago without federal review. In addition, ecological impacts, environmental mitigation, and simple scale of the road surface area vary widely by road type. For example, depending on the scale of concern, an eight-lane interstate highway connecting major cities would have much greater fragmenting effects than a two-lane rural road.

## Early Roads and Turnpikes

Early colonial routes were mostly natural surfaces intended to allow for the passage of wagons. These roads were built mainly to complement an extensive waterway transportation system. Roads provided local access and allowed the movement of people and goods where canals or other water courses were not viable. The roads and waterways rarely competed with one another.

The federal government became involved in road construction to develop interior lands and a national postal service and to defend remote territories. Settlers purchasing land from the government generated revenue for the federal government to build roads. This revenue was an important source for road building and maintenance, especially in sparsely populated areas.

In 1806, the federal government began its most ambitious project to date: construction of the National Road, also known as the Cumberland Road, from the Potomac River at Cumberland, Maryland, inland to the Ohio River at Wheeling, West Virginia. By 1831, revenue began drying up, completed sections of the Cumberland Road were turned over to the states, and the federal government halted all road funding. Local governments, however, oversaw all road construction and operation from the 1830s to the 1920s because the states had little interest in these activities. A major growth in immigration spurred western migration, and by the late 1880s, there were many established routes within and between cities, as well as established routes for interstate and transcontinental travel. Interest in improving roads began again in the late 1800s as bicycles proliferated and in the early twentieth century as cars became more common (Forman et al. 2003).

The first response to improving roads was oiling of the naturally surfaced roads. Oiling was followed by paving, using an asphalt surface. Most of the paving was done on a small scale until World War I. After the war, the federal government greatly increased paving and road building to "get the farmer out of the mud." The funding mechanism for the improvements was a tax on gasoline sales. By 1923, 33 states had imposed a gas tax, and by 1929, all states had imposed the tax. The federal government enacted the first national gas tax in 1932, later converting the tax on gasoline and other transport-related products, such as tires, into a dedicated trust fund for highways (Forman et al. 2003).

## Federal Aid Highways

In 1893, with the creation of the Office of Road Inquiry in the U.S. Department of Agriculture, the federal government renewed its interest in road development. The first serious effort was in 1916 when the U.S.

Congress created the Federal-Aid Highway Program, which continues as the basic program for federal support of highways. The Federal-Aid Highway Program specified that states receive federal funds subject to various conditions, one of which required the states to match federal funds dollar for dollar. The federal program also required states to create a state highway department that was technically oriented and managed in accordance with the principles of scientific management and administration. The state-level department was meant to serve as a partner in the federal program and was given sufficient authority to supervise the expenditure of the funds. Federal funds could not be passed to localities. Reliance on the states as the senior partners in the Federal-Aid Highway Program continues to the present, although local officials in counties and municipalities have played a stronger role since the early 1960s (Forman et al. 2003).

## Interstate Highway System

A few intercity highways, such as the Merritt Parkway in Connecticut, and the New Jersey and Pennsylvania turnpikes were built in the 1930s through the 1950s, and a few major cities had some limited-access, divided, multilane highways. There was no national system of freeways, however, until 1956, when the U.S. Congress enacted a plan to build and finance the National System of Interstate and Defense Highways, now known as the interstate highway system, to serve auto, truck, and strategic military needs. The interstate system was to be 42,500 miles of four-lane divided highways with limited access throughout. Standard vertical and horizontal clearances were designed to support military vehicles, such as trucks carrying tanks. The federal government would pay 90% of the cost (Forman et al. 2003).

The interstate highway system was considered complete in 1990 and could be enlarged only if a state used its own funds to build a road to interstate standards and then petitioned the federal government to have the route added. The post-interstate era began in 1991 with the passing of the Intermodal Surface Transportation Efficiency Act (ISTEA). This act and the following reauthorization are discussed in Chapter 5.

For the past 20 years or so, the funding and management of the interstate highway system and the national highway system (NHS) are distinct, but their environmental approaches are similar.

## Present U.S. Road Network

There are approximately 4 million miles of public roads in the United States and 8.3 million lane miles. The majority (76% of lane miles) of paved public roads in the United States are two-lane rural highways, and the remainder are urban and rural multilane roads. The road network is expanding slowly, having only 55,000 lane miles built between 1987 and 1997 (less than 0.2% increase per year). About 80% of the expansion to the system is from road widening. As discussed in Chapters 4 and 6, this mode of growth has implications for the types of methods and the data used for assessment and management of ecological impacts.

## DEFINITIONS AND CHARACTERISTICS OF ROADS

### Vehicle Miles of Travel

The annual vehicle miles of travel (VMT) is a measure of the demand and use on the road system. Statistics such as annual VMT help transportation agencies plan for the future. The amount of VMT has been steadily increasing (Figure 2-1). By 2000, annual travel on the nation's highways reached about four times the travel level in 1960, or an estimated 2.7 trillion VMT.

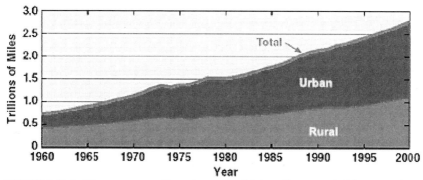

**FIGURE 2-1** Time course of number of vehicle miles traveled by year from 1960 to 2000, indicating travel on rural and urban portions of the highway system. Source: FHWA 2001b.

Road use is increasingly made up of urban traffic. At some point in the 1970s, total VMT on urban roads exceeded that on rural roads (Figure 2-1). Urban roads and streets now carry about 1.7 trillion VMT, or about 61% of total VMT. Data from the past decade indicate that the rate of road use increased more rapidly than the rate of system growth in length. From 1990 to 2000, travel increased 28.9%, yet the total miles of roads in the United States increased by only 2.1%. The increased volume is attributed in part to the increasing urbanization of the country. As urban populations increased and urban boundaries expanded, urban travel increased 30.6%, and rural travel increased 24.9% from 1990 to 2000. During the 1990s, the largest increase (41.9%) in road use occurred on the principal arterial urban roads (not urban interstates) (Table 2-1).

## Road Density

Less than half of 1% of the land area in the United States is covered by roads of all kinds, not including rights-of-way, parking lots, and driveways. The United States has 1.2 miles of roads for every square mile of land area, much less than many other developed nations. Japan has a road density approximately 4 times greater than that of the United States. Germany, France, and England have road densities 2.5 times greater than that of the United States, and those densities continue to grow even in Europe's high-density environments. However, Canada has a density of only about 0.16 miles per square mile of land. Densities also vary greatly across the United States, the highest being in New Jersey and the lowest in Alaska.

## Road Surfaces

Of the 4 million miles of roads in the United States, about 2.3 million miles (59%) are paved. There are two general classes of materials used for construction of pavement: flexible pavements (asphalt) and rigid (hydraulic cement concrete) pavements (AASHTO 1993, http://www.fhwa.dot.gov/infrastructure/materialsgrp/cement.html). Flexible pavements generally consist of a prepared roadbed or subgrade, subbase, base, and surface layer. In some cases, the base is used as a drainage layer, or the drainage layer may be part of or below the subbase. In contrast, rigid pavements have

**TABLE 2-1** Functional System Changes, 1990-2000

| Functional System | Rural VMT | % Change 1990-2000 | Urban VMT | % Change 1990-2000 | Total VMT | % Change 1990-2000 | % of Total Travel |
|---|---|---|---|---|---|---|---|
| Interstate | 270,315 | 34.5 | 397,288 | 41.3 | 667,603 | 39.4 | 24.1 |
| Freeways/ expressways | — | — | 178,105 | 38.6 | 178,105 | 38.6 | 6.4 |
| Principal arterial | 249,137 | 41.9 | 401,237 | 18.9 | 650,374 | 27.4 | 23.5 |
| Minor arterial | 172,780 | 10.5 | 326,855 | 37.5 | 499,635 | 27.5 | 18.1 |
| Major collector | 210,496 | 9.9 | — | — | 210,496 | 9.9 | 7.6 |
| Minor collector | 58,571 | 16.3 | — | — | 58,571 | 16.3 | 2.1 |
| Collector | — | — | 137,008 | 27.5 | 137,008 | 27.5 | 5 |
| Local | 128,332 | 31 | 237,239 | 23.5 | 365,571 | 26.7 | 13.2 |
| Total | 1,089,631 | 24.9 | 1,677,732 | 30.6 | 2,767,363 | 28.9 | 100 |

Source: FHWA 2001b.

only a subbase between the prepared roadbed and pavement slab. Another distinguishing characteristic of flexible versus rigid pavements is the distribution of traffic load over the prepared roadbed. Rigid pavements tend to distribute the traffic load over a wide area, and flexible pavements, made with pliable material, do not spread loads as widely and thus usually require the additional base layer and greater thickness for optimal traffic-load transmittal. Of the paved roads, approximately 95% use flexible pavement, and 5% use rigid pavement.

## Types of Roads

The current road system consists of several types of roads placed into a functional class based on traffic volume and general use. Unless otherwise stated, the following information was taken from the Federal Highway Administration (FHWA 2001b).

Roads are classified on the basis of the types of function provided. Functional types of roads that constitute the highway network include (1) interstate highways, (2) arterials, (3) collectors, and (4) local streets or access roads. These four major types of roads can be combined into groupings for administrative and funding purposes. One major grouping is the NHS, which comprises the interstate highways and a large number of the high-volume arterial roads. Although not exclusive, most of the considerations of ecological impacts that occur in subsequent chapters consider the impacts from the first three functional types listed above. As discussed in Chapters 3 and 4, ecological considerations such as environmental mitigation and simple physical scale vary by road type.

Recognizing that a road does not serve traffic needs by itself is basic to the development of any logical highway system. Travel involves movement through a network of interrelated roads and streets. The movement channels through an efficient hierarchical system that includes lower-order roads that handle short and local trips to higher-order roads that connect regional and interregional traffic and longer trips. In addition to movement, access is a fundamental function of roads. The principal function of local roads, almost 70% of the total mileage, is access, whereas freeways limit access.

Federal law requires functional designations of roads in urban and rural areas for funding purposes. This classification is done by state transportation agencies and is mapped and submitted to Federal Highway Administration (FHWA) to serve as the official record for the federal highway system. The distinction between rural and urban areas is made using federal census data to create federal-aid and urban-area boundaries (WSDOT 2003). Urban roads occur in a census area with an urban population of 5,000 to 49,000 or in a designated urban area with a population greater than 50,000. Rural roads are defined as any road not located within the urban-area boundary.

## High-Speed Limited-Access Highways

The best known high-speed limited-access highway system is the current interstate highway system. Because of the need to accommodate heavy freight traffic, these roads are the most expensive to build and maintain.

Other types of limited-access highways include some roads within the NHS or some state limited-access roads, such as the New York Thruway. These limited-access state highways are funded and administered under their own sets of legal standards. Despite important administrative differences, the committee determined that the differences in ecological impacts of different types of limited-access highways are minor.

The interstate system accounts for only 1.2% of the nation's total miles of roadway (Table 2-2), yet interstate highways convey 24% of the annual VMT in the United States and 41% of total truck VMT, suggesting their importance to commercial transportation. These statistics also indicate the potential for the interstate system to deliver greater levels of contaminants to air and water than the total miles of interstate roadway would suggest. However, other factors suggest that the contaminant load is not simply proportional to VMT. For example, vehicles traveling at interstate speeds may emit some pollutants at a lower rate than vehicles operating on local streets.

**TABLE 2-2** Interstate Highway System—Key Statistics

|  | Interstate System | Total Highway System | Interstate System Share (%) |
|---|---|---|---|
| Road miles | 46,677 | 3,951,098 | 1.2 |
| Lane miles | 209,655 | 8,328,856 | 2.5 |
| VMT (billion/yr) | 667 | 2,767 | 24.1 |

Source: AASHTO 2002a. Reprinted with permission; copyright 2002, American Association of State Highway and Transportation Officials, Washington, DC.

Other federal highway investment, in addition to the interstate highway system, is reflected in components of the NHS. The NHS consists of the many routes with such designations as U.S. 1 or U.S. 89. These roads may be high-speed limited-access highways, arterials, or collectors, depending on the location and configuration of the roadway. The NHS includes urban and rural roads that serve a wide variety of transportation functions.

The NHS comprises 163,000 miles of roads, 46,477 of which are interstate highways. The NHS serves major population centers, intermodal transportation facilities, international border crossings, and major travel destinations. It includes the interstate systems, other rural and urban principal arterials, highways that provide access to major intermodal transportation facilities, strategic highway network connectors, and the defense strategic highway network (Figure 2-2). The NHS carries over 44% of the nation's travel, but it makes up only 4% of the nation's total public roadway miles (Table 2-3). Although more than 70% of NHS miles are in rural areas, almost 60% of VMT on the NHS take place in urban areas (Figure 2-3).

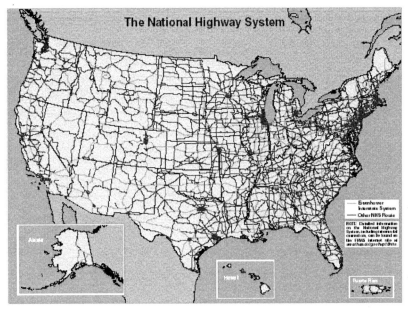

**FIGURE 2-2** Map of National Highway System in the 50 states, the District of Columbia, and Puerto Rico. Source: FHWA 2001b.

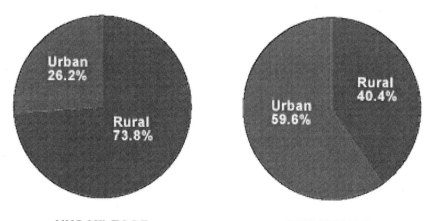

**NHS MILEAGE**                    **NHS TRAVEL**

**FIGURE 2-3** Miles of road and use of roads in urban and rural portions of the National Highway System. Source: FHWA 2001b.

**TABLE 2-3** National Highway System Mileage and Travel in Rural and Urban Areas

|  | Rural | Urban | Total |
|---|---|---|---|
| NHS Mileage |  |  |  |
| Interstate | 33,150 | 13,527 | 46,667 |
| Other NHS | 85,882 | 28,629 | 114,511 |
| Total NHS | 119,032 | 42,156 | 161,188 |
| NHS Percent of Total Mileage |  |  |  |
| Interstate | 0.8 | 0.3 | 1.2 |
| Other NHS | 2.2 | 0.7 | 2.9 |
| Total NHS | 3.0 | 1.1 | 4.1 |
| NHS Travel (millions VMT/yr) |  |  |  |
| Interstate | 270,315 | 397,288 | 667,603 |
| Other NHS | 224,340 | 333,335 | 557,675 |
| Total NHS | 494,655 | 730,623 | 1,225,278 |
| NHS Percent of Total Travel |  |  |  |
| Interstate | 9.8 | 14.4 | 24.1 |
| Other NHS | 8.1 | 12.0 | 20.2 |
| Total NHS | 17.9 | 26.4 | 44.3 |

Source: FHWA 2001b.

## Arterials

Arterials consist of the interstate highway system, multilane limited-access freeways and expressways, and other road corridors that serve local areas; they also carry substantial statewide or interstate travel volumes. This system accounts for approximately 11.1% of the nation's public roadway mileage and carries approximately 72.1% of the total VMT (Figure 2-4).

## Collectors

Collectors connect local streets and roads with arterials. They provide traffic circulation and land access among downtown city centers, industrial and commercial areas, and residential neighborhoods. Collectors provide lower speeds and less mobility for shorter distances than arterials.

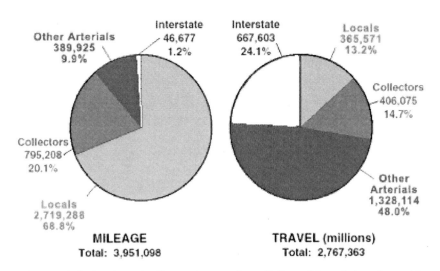

**FIGURE 2-4** Total road mileage (summation of all road lengths) and travel by functional type in the year 2000. Travel represents vehicle miles traveled per year. Source: FHWA 2001.

## Locals

The roads and streets functionally classified as local are all those not classified as part of the principal arterial, minor arterial, and collector system. Local roads and streets primarily provide access to adjacent land and are generally not maintained by a state highway agency. Local functional systems serve only 13.2% of total VMT, but these roads account for 68.8% of the nation's total roadway miles (Figure 2-4).

## Layout Patterns

The growth of suburbs, and the spread-out development patterns that come with it, has had a major influence on urban roadway patterns beginning in the 1920s but particularly occurring after World War II. The layout of the road system is not only a function of history and geography but is also linked to the local and regional development patterns. Layout patterns give roads different types of connectivity. The layout patterns and spacing among roads as well as their width (number of lanes) determine road density. There are several types of roadway layouts that are being used for designing roads. These patterns can vary greatly from city to city but generally involve a rectangular grid, a hub-and-spoke layout, or a combination of the two.

## Rectangular Grids

The road network in most cities across the United States is an outgrowth of the Public Land Surveys and follows a rectangular grid pattern (Figure 2-5). Roads are orthogonal, alignment generally being along ordinal directions (north-south or east-west). Grid patterns generally have more total street length, blocks, intersections, and access points than other layouts. The grid pattern is typically intertwined with a mixed pattern of land use. Mixed land use occurs when different functions of use (residential, commercial, and industrial) are not geographically separated, but "mixed" in the landscape. Mixed land use was the dominant development style in American cities and towns in the early twentieth century and continued to be the primary pattern until the development of suburbs after World War II. A mixed land-use pattern can ease conges-

**FIGURE 2-5** Example of rectangular grid pattern of roads in Chicago. Source: MapQuest 2005. Reprinted with permission; copyright 2005, MapQuest.com, Inc., and GDT, Inc. The MapQuest.com logo is a registered trademark of MapQuest.com, Inc.

tion on main streets by offering acceptable alternative routes. It may also reduce the total VMT if housing and services are intermixed. However, it can also add unwanted through traffic on some residential streets (Berkovitz 2001).

### Hub-and-Spoke Patterns

In many locations, the hub-and-spoke pattern developed with the local growth pattern in the 1960s and 1970s. This system of roads comprises circular roads (hubs, belts, and ring roads) that go around a city center at various distances and separate roads (spokes) that go to the center of town (Figure 2-6). The circular roads are often used to route traffic around major urban areas, and the spokes are designed for commuter traffic. Freeways and expressways (interstate) often form the circular hubs and major spokes in urban and suburban areas.

The hub-and-spoke pattern is correlated with a type of land-use pattern that is designated as conventional land use. This pattern arose, in part, because of central urban planning and the increased importance of

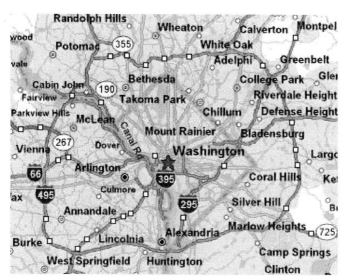

**FIGURE 2-6** Example of hub-and-spoke pattern of roads found in Washington, DC. Source: MapQuest 2005. Source: MapQuest 2005. Reprinted with permission; copyright 2005, MapQuest.com, Inc., and GDT, Inc. The MapQuest.com logo is a registered trademark of MapQuest.com, Inc.

the automobile in everyday life. Suburban areas grew in number and size to house families.

Residential areas are often built with cul de sacs and a small number of entry points to reduce the amount of pass-by traffic in residential neighborhoods (Berkovitz 2001). In the conventional pattern of land use, each type of land use (residential, commercial, retail, and industrial) is separated from the others. Conventional land-use patterns result in more of a hub-and-spoke or circulatory pattern, with businesses in the center of town and residential areas surrounding the city. Some of the longest commutes in metropolitan regions are made by residents who live at the metropolitan edge and who work in downtown areas (FHWA 2002b).

## Engineering Structures

In addition to roads and roadsides, the road system includes many engineering structures. These include concrete barriers, guardrails, noise barriers, bridges, culverts and pipes, and overpasses and underpasses. Each of these structures has a particular ecological effect.

## Concrete Barriers

Rigid safety barriers separating lanes on roads are common, especially in urban areas. The most common type of these structures is called a Jersey barrier. Some Jersey barriers are used for traffic separation on freeways and interstate highways, and many others are used only temporarily. Temporary units are used mainly to enhance safety in construction work zones. The height of Jersey barriers averages around 32 in., but some are as high as 57 in.

Rigid safety barriers are also commonly used longitudinally on major nonlimited-access highways to safely separate the two directions of traffic and preclude left turns and U-turns. There is current concern with barriers being formidable obstacles to small- and large-animal movement across highways. Barriers can block animals when they attempt to cross the highway, making them vulnerable to traffic mortality.

## Guardrail and Right-of-Way Fences

Many different types of horizontal flexible and semirigid barriers, commonly known as guardrails, line the roadsides and medians of the roadway system. As with concrete barriers, guardrails are intended to constrain errant cars and trucks. Guardrails normally provide space beneath and above the rail that allows movement of wildlife. The principal difference between rigid safety barriers and guardrails is the amount deflection that will occur when they are hit.

Right-of-way fences are used to keep people and animals from entering the berm, shoulders, and travel lanes of the highway. Many limited-access highways are lined with right-of-way fences, which enhance safety and reduce traffic mortality of large animals by reducing the crossing of animals.

## Wide Medians

American Association of State Highway and Transportation Officials (AASHTO) safety information reflects a continuum of preference among lane-separation technologies. The safest system is one with wide grassy medians and no hard barriers, where there is sufficient room for errant vehicles to recover before entering an opposing lane of traffic. No rigid barrier is needed if the median is approximately 50 ft or more in width. Medians less than 30 ft in width should have a rigid barrier (AASHTO 2002a, TRB 2003, Mak and Sicking 2003).

Despite the safety offered by wide medians, there can be offsetting concerns because the median increases the width of the road right-of-way, including additional cost, availability of land in some locations, and compatibility of a wide right-of-way with nearby land uses.

Although wide medians in an additional loss of natural habitat, they may have less ecological impact than a narrow, paved median with a barrier or guardrail. Another advantage of a wide median is that the highway can be cost-effectively widened in the future with minimal disruption to traffic and to the adjacent properties, although if that happened, the ecological benefits of a wide median would be minimized or eliminated.

## Overpasses and Underpasses

The decision to construct a new road over or under an existing facility is site-specific. Generally, minor roads should pass over major roads. This configuration takes advantage of off-ramp traffic being able to decelerate on the upgrade and the on-ramp traffic being able to accelerate on the downgrade (Sharpe 2004). Although highways generally are constructed over waterways, the choice to span or pass under requires extensive analysis. Sometimes tunnels are built instead of bridges, as was the case with the 2.75-km Fort McHenry tunnel, the largest underwater highway and widest vehicular tunnel, submerged in a trench at the bottom of the Patapsco River in Baltimore, Maryland. A tunnel was constructed instead of a bridge because of concerns that a bridge would tower over a historic site, Fort McHenry.

Often it is desirable to depress the major road to reduce noise impacts and improve aesthetics. Span lengths, angle of skew, soil conditions, drainage, and the maintenance and protection of traffic on the existing route all must be considered. Three-dimensional models of the alternatives are sometimes used to obtain informed public input into the decision.

## Noise Barriers

Noise barriers are designed and built primarily to muffle highway traffic noise in residential areas, schools, playgrounds, and other sensitive receptor areas (see Chapter 5 for a discussion on noise standards). Noise barriers may also be retrofitted on highways to enhance the surrounding community. Almost all sound walls in the United States have been constructed since 1986, and about two-thirds have been constructed of precast concrete and block. The other one-third of barriers are constructed of various materials, including wood, earth berm, and metal. The average noise barrier height is 10 to 16 ft, although some are over 23 ft.

### Bridges, Culverts, and Pipes

The road system in the United States includes over 580,000 bridges, of which approximately 481,000 (83%) are over waterways. Ongoing construction efforts to address deficient bridges offer a singular opportunity to address and mitigate important ecological issues. These issues are discussed in Chapter 3; mitigation of adverse effects of bridges and culverts is discussed in Chapter 4. The condition of bridges and other engineering structures is discussed later in this chapter.

### OWNERSHIP AND MAINTENANCE RESPONSIBILITIES

Understanding the ownership and maintenance responsibilities for roadways is important for planning and managing the coordination of environmental and planning issues discussed in later chapters of this report. Local town, city, and county governments own and maintain 77.4% of the nation's roads. The federal government owns only 3.0% of the roads, including those in national forests, parks, military areas, and Indian reservations. Thus, coordination of transportation and environmental issues (including ecological protection) is often a state or local planning concern. Individual states own the remaining 19.6% of the roads, which includes most of the interstate system (Table 2-4).

### CURRENT STATUS AND FUTURE TRENDS

In 2000, FHWA (2001b) assessed the status of the U.S. road system. It reported that most of the existing road surfaces in the system are in good condition but that an aging set of bridges and other engineering structures are a challenge to the system. The system is carrying higher

**TABLE 2-4** Ownership of U.S. Roads and Streets

| Jurisdiction | Rural Mileage | % | Urban Mileage | % | Total Mileage | % |
|---|---|---|---|---|---|---|
| State | 663,755 | 21.5 | 111,539 | 13 | 775,294 | 19.6 |
| Local | 2,311,269 | 74.7 | 746,341 | 86. | 3,057,610 | 77.4 |
| Federal | 116,724 | 3.8 | 1,484 | 0.2 | 118,208 | 3.0 |
| Total | 3,091,748 | 100.0 | 859,364 | 100 | 3,951,112 | 100.0 |

Source: FHWA 2001b.

volumes of traffic than in the past. Many roads are approaching design capacity as congestion increases. These issues are discussed in the following sections.

### Travel Congestion on Principal Arterial Roads in Urban Areas

The use of urban roads continues to outpace the use of rural ones. The use and subsequent congestion are evident on both interstate and principal arterial roads. Although interstates play a key role in intercity connectivity, military support, and efficient long-distance travel, they also support local travel. Commercial uses (moving freight within metropolitan areas), noncommercial uses (providing access to airports and commuter travel), and individual uses are increasing the use of urban interstates. Reflecting the evolving expectations that state and local officials have for the interstates, growth in travel on their urban segments has been greater than on the rural portions, a trend that has continued since 1960 (Figure 2-6; FHWA 2001b). Travel per lane mile in rural areas has more than doubled since 1960. Travel per lane mile in urban areas has increased even more, growing roughly 25% during the 1990s. Urban interstate congestion is high in nearly half the states. Forty-one states in a recent General Accounting Office survey predicted that congestion would be high or very high in a decade (AASHTO 2002a).

The traffic volume-to-service flow (capacity) ratio (V/SF) is one measure of congestion. As the ratio approaches the theoretical value of 1.00 (volume of traffic is equal to service flow capacity of the facility), traffic slows down and eventually stops. Values of V/SF greater than or equal to 0.80 indicate congestion. Level of service (LOS) is another concept used in transportation work to describe different levels of congestion. LOS is determined by comparing the volume of traffic on a road section to the road's capacity to handle the volume. LOS A is free-running traffic, and LOS F is gridlock (Table 2-5 and Figure 2-7). Previously, FHWA and state Departments of Transportation would try to attain LOS C or better during the peak hours; now, due to heavy travel demand, they have to accept LOS D or even LOS E as goals during rush hour.

Many sources suggest that travel is becoming more congested. The percentage of travel under congested conditions (percentage of daily traffic on freeways and principal arterial streets in urban areas) has slowly increased from just over 26% in 1987 to over 33% in 2000 (Figure 2-8). However, off-peak travel has increased significantly in most major metropolitan highway systems in the past decade. Urban interstates are the most crowded. In 2000, 50% of peak-hour travel occurred under congested conditions (FHWA 2002b). Researchers estimated that the annual costs due to lost productivity from congestion-related delays were $78 billion in 1999 (FHWA 2002b).

## Pavement Condition

Over the past several years, the overall condition of the pavement on the NHS and the interstate system has improved. In 2000, on the basis of the international roughness index (IRI), 93.5% of the NHS and 96.6% of the interstate system had acceptable ride qualities. The IRI is an indicator of pavement performance, which uses an objective instrument-based rating system to measure the ride quality of a road. Pavements with an IRI of less than 95 have good or very good ride quality, those with an IRI of 95-170 have an acceptable ride quality, and those with an IRI of more than 170 have an unacceptable ride quality. The NHS and the interstate system have a 6.5% and 3.4%, respectively, unacceptable ride quality (Figure 2-9).

**TABLE 2-5** Level-of-Service Chart for Major State Highways in New Hampshire Based on 2002 Traffic Data

| Level of Service Description | Miles |
|---|---|
| No Congestion (LOS A and B) | 1,233 |
| Moderate congestion (LOS C and D) | 1,191 |
| Congested (LOS E and F) | 305 |
| Total | 2,729 |

Source: Modified from NHDOT 2004.

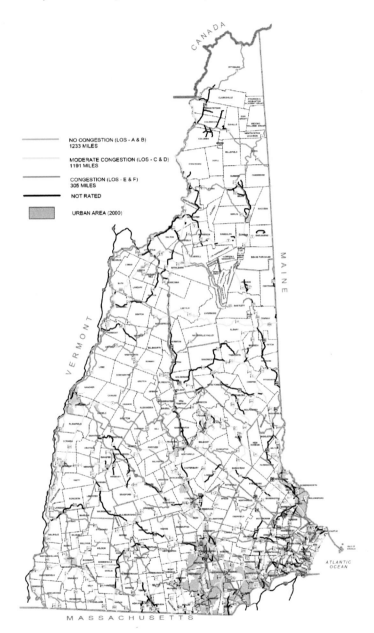

**FIGURE 2-7** Traffic congestion map showing a concentration of congested highways in the southeastern and south-central regions of New Hampshire. Green: no congestion (LOS A and B); yellow: moderate congestion (LOS C and D); and red: congested (LOS E and F). Source: NHDOT 2004.

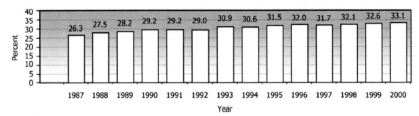

**FIGURE 2-8** Percentage of vehicle miles traveled on urban freeways and principal arterials occurring under congested conditions from 1987 to 2000. Source: Schrank and Lomax 2001. Reprinted with permission; copyright 2001, Texas Transportation Institute.

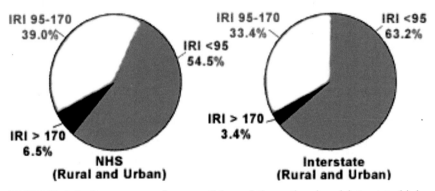

**FIGURE 2-9** Pavement surface condition of the national and interstate highway systems. Source: FHWA 2001b.

The condition of interstate pavement has improved in recent years, reflecting greater federal funding and the states' commitment to maintaining these roads. Because of increasingly heavy use, most of the system requires substantial maintenance.

Porous pavement is a newer technology for the primary highway system. Structurally, porous pavement has not proved to be as durable to heavy trucks and high traffic volumes as other road surfaces because of movement of water under the road, resulting in cracking. Porous pavement also is a problem in states having to deal with snow removal because the sand and salts used in snow removal clog the openings. The most promising application for porous pavements is in parking lots, but even most of these fail over time as a result of compaction.

## Bridge Conditions

Many of the nation's bridges and road structures (overpasses and underpasses) were built many years ago and are in need of maintenance and, in many cases, rehabilitation. These features, which represented considerable expense at the time of design and construction, also require ongoing maintenance.

In 2000, 29% of the nation's estimated 590,000 bridges were considered structurally deficient or functionally obsolete (Table 2-6). These categories are defined below.

### Structurally Deficient

Structurally deficient bridges have deteriorated structural components and are usually closed or restricted to use only by light-duty vehicles. These bridges are not necessarily unsafe. In general, strict adherence to signs limiting speeds or quantity of traffic on bridges will provide sufficient safeguards for these bridges.

### Functionally Obsolete

About half of the nation's 590,000 bridges were built before 1965, and a fourth are more than 50 years old. Although bridges that are properly cared for can be considered virtually permanent, their age means that they need substantial maintenance and may be functionally obsolete (AASHTO 2002a).

**TABLE 2-6** Conditions of Highway Bridges

| Condition | Total Highways | |
|---|---|---|
| | No. | % |
| Structurally deficient | 88,150 | 15.0 |
| Functionally obsolete | 81,900 | 14.0 |
| Acceptable | 415,492 | 71.0 |
| Total | 585,542 | 100 |

Source: FHWA 2001b.

Functionally obsolete bridges cannot safely support the type or volume of traffic using them. These bridges have older design features that limit them from carrying current traffic volumes and modern vehicle weights and sizes. They can also cause more environmental damage, such as alteration of floodplain and riparian habitats, because of water flow restrictions.

Investments made possible by ISTEA and the Transportation Equity Act for the Twenty-First Century (TEA-21) have permitted bridge improvements. The investment needed to repair the backlog of bridges has decreased in tandem with the reduction in bridge deficiencies. The bridge investment backlog is now $52 billion, based on an evaluation using the new National Bridge Investment Analysis System. Table 2-7 below shows the breakdown on bridge needs. Progress has been made through increased funding under ISTEA and TEA-21, priorities set at the state and other levels, and reduced investment requirements that meet economic analysis criteria.

## SUMMARY

This chapter provides an overview of the status of the U.S. road system. The perspective has been on the entire road system at the broadest spatial scales. The spatial template for the system is a combination of established use histories (following older, established routes) and planning. The network comprises different functional types of roads (interstate, arterial, connector, and access) as well as rural or urban areas where roads traverse. The functional roadway types are used for planning and funding purposes. The characteristics of the current U.S. road system indicate that most of the growth in the paved system now occurs via widening existing roadways rather than adding new routes to the system.

**TABLE 2-7** Backlog of Bridge Investment Needs

| Type of Investment | Costs (Billions) |
| --- | --- |
| Bridge replacement needs | $37.2 |
| Bridge improvement needs (widening, raising, strengthening) | $3.1 |
| Maintenance, rehabilitation, and reconstruction needs | $11.6 |
| Total | $51.6 |

Source: AASHTO 2002a.

Vehicle use of roads has steadily increased in the past century. Since the 1960s, urban road travel has become greater than rural road travel; yet the later has more than doubled, and that has important ecological impacts, as discussed in Chapter 3. Indicators of congestion in urban areas are increasing. The extant road system requires large amounts of maintenance and replacement to maintain its current function. Although paved sections of the NHS are considered to be in very good condition for driving, 3 of 10 bridges are structurally deficient or obsolete.

# 3

# Effects of Roads on Ecological Conditions

## INTRODUCTION

Widespread attention continues to be drawn to the ecological effects of roads, especially as the road system continues to expand. Roads are created because of changing interactions between people and their environments. They are created to facilitate access to natural resources, to connect human communities, to move goods to markets, and to move people to work. Whatever their purpose, roads, road establishment, road maintenance, and road travel have a broad variety of effects.

In this chapter, the committee addresses the following two questions as stated in its charge:

1. What are the appropriate spatial scales for different ecological processes that might be affected by roads?
2. What are the effects of road density on ecosystem structure and functioning and on the provision of ecosystem goods and services?

Roads have effects that can vary with a range of spatial scales. The committee's analysis examines what is known about road effects at three scales, which are discussed later in the chapter.

For the second question, the committee used the phrase "ecological condition" for "ecosystem structure and functioning." Because most information on road effects is given in terms of ecological structure and functioning rather than ecosystem goods and services, ecological condition is used. These terms are defined in the next section.

This chapter is organized into five sections to summarize the interaction between roads and ecological conditions. After this introduction, terms and concepts are defined in the second section. A short summary of effects on ecological goods and services is provided in the third section. The fourth section summarizes published, mostly refereed literature and is organized according to two dimensions: the first is the ecological process of interest (many of which are listed in Table 3-1) and includes the effects of each process on the different levels of ecological organization (for example, abiotic, population, species, and ecosystem), and the second is the scale of the effect. Many of the effects of roads on the environment are caused by other forms of human activity and land use. Impacts of agriculture, urbanization, forest practices, and manufacturing are in many ways similar and sometimes interrelated with the impacts of roads. Information gaps are discussed in the fifth section.

## DEFINITIONS

*Ecological condition* is a general term that describes the structure and functioning of ecosystems. It may refer to the status of the ecological environment at a particular time or to dynamic changes in its components and processes over time. The dynamic aspects of these condition measurements are discussed later in the chapter where both spatial and temporal dimensions are addressed. Ecosystems encompass all living organisms (*biotic components*) plus the nonliving environments (*abiotic components*) with which they interact. The abiotic components consist of hydrological and geomorphological processes, chemicals, and such disturbances as landslides, climate and weather. Levels of organization of biotic components used in this report are genetics, species and population (plants and animals), and ecosystem. Each level of biotic components has attributes of composition, structure, and functioning, and together constitute biological diversity (often called "biodiversity").

*Composition* refers to the identity and variety of elements in each of the biodiversity components. *Structure* refers to the physical organization or pattern of the elements. *Ecological (or ecosystem) functioning* refers to the ecological and evolutionary processes acting among the elements, or how the ecosystem works.

**TABLE 3-1** Comparison of Ecosystem Goods and Services[a] and Ecosystem Structures and Processes Affected by Roads

| Change Due to Roads | Consequence | Affected Ecosystem Good | Affected Ecosystem Service |
|---|---|---|---|
| Chemical input from roads to water bodies | Degradation of water quality, bioaccumulation | Clean water | Water purification, pollution abatement |
| Chemical inputs to airshed | Degradation of air quality | Clean air | Pollution abatement |
| Chemical input to soils | Bioaccumulation | Soil fertility | Pollution abatement |
| Climate | Increased temperature and rainfall | Water | Climate stability |
| Hydrological processes | Fluvial dynamics, sediment transport, floodplain ecology | NA | Flood and drought mitigation, nutrient cycling |
| Modified habitat | Plant species composition (natives and nonnatives) | Biodiversity | Nutrient cycling, soil fertility, seed dispersal |
| Habitat quality, wildlife mortality | Density and composition of animal species and populations | Biodiversity | Crop pollination, aesthetics, ecotourism |

[a]Ecosystem goods and services are defined by Daily (1997).
Abbreviation: NA, not available.

*Diversity of the genetic component* refers to the variation in genes within a particular species, subspecies, or population. A relevant measure of the genetic component is allelic diversity, a measure of its structure is heterozygosity, and a measure of its functioning is gene flow.

*Diversity of the population and species component* refers to the variety of living species and their populations at the local, regional, or global scale. A relevant measure of this component is species abundance, a measure of its structure is population age, and a measure of its functioning is demographic processes, such as births and deaths.

*Diversity of the ecosystem component* refers to the number of species, biotic communities, and ecosystem types and to the genetic variation in the organisms present. Relevant measures of this component include a variety of measures of species diversity (see NRC 2001), including the ratio of native to nonnative species; various measures of genetic diversity (see Hedrick 2004); and variations in trophic activities and structure, such as food-chain length and feeding adaptations.

## ECOLOGICAL CONDITION—
## ECOSYSTEM GOODS AND SERVICES

Ecological condition incorporates the concepts of ecosystem goods and services through the ecosystem component. Ecosystem goods are the materials and elements (for example, water, food, fiber, and fuel) that are products of ecosystems and used for a variety of human needs. Ecosystem services are the benefits that people obtain from ecosystems. Those include goods, such as food and water; services, such as regulation of floods, droughts, land degradation, and disease; supporting services, such as soil formation and nutrient cycling; and cultural goods, such as recreational, spiritual, religious, and other nonmaterial benefits (Millennium Ecosystem Assessment 2003).

An enumeration of road effects on ecosystem goods and services is marginally addressed in this report. The effects of roads on those ecological structures and processes that are of direct and indirect use to humans are, however, discussed in detail. Ecosystem structure and functioning can be translated into ecosystem goods and services, as described in subsequent sections.

Roads affect ecosystem goods and services in many ways. The documented road-associated changes to be discussed can be translated into equivalent alterations in ecosystem goods and services (Table 3-1).

Among the changes are altered local climatic conditions, altered nutrient cycling, loss of flood and drought mitigation, loss of soil fertility, and changes in biodiversity.

## Ecological Conditions and Scale

Roads interact with ecosystems across a wide range of scales. For example, at small scales, heavy metal molecules accumulate in soils adjacent to roads. At intermediate scales, roads disrupt soil structures and hydrological pathways and alter plant and animal communities. At large scales (regions to nation), roads alter migration patterns and increase spread of exotic organisms. Many effects can occur at more than one spatial scale (for example, effects on migration patterns). The literature review in the next section documents effects of roads on ecological conditions at three scales. The smallest scale assessed is the road segment. This scale generally extends to a hundred meters (see Figures 1-1, 1-2, and 1-3). The intermediate scale is identified as a system of a geographic region—defined either politically (for example, a state or province) or ecologically (for example, a watershed or eco-region). This scale extends from one to tens of kilometers (see Figures 1-4 and 1-5). The largest scale is the macroscale and is defined by regional ecological units (for example, eco-regions) (Bailey et al. 1994) to national political boundaries. This scale extends from hundreds to thousands of kilometers (see Figures 1-6 and 1-7).

## LITERATURE REVIEW

The committee developed an annotated bibliography (Appendix B) of road effects on ecological conditions, with an emphasis on spatial scale. The review included only studies that directly measured the effects of roads on the surrounding environment. Effects were categorized as either abiotic or biotic. Abiotic effects included the effects on hydro-geomorphic process, the effects of road-related chemicals on water and air quality, and the effects of other disturbances, such as landslides, local climate, and lighting. The three subcategories of biotic effects were genetic, species and population, and ecosystem. Within each subcategory, the effects of roads on structure, functioning, and composition were documented.

Every aspect of roads (as with many human activities) has some interaction with the surrounding environment, including road construction, operation, and maintenance. However, the committee's review focuses on operation—that is, the effects of roads and their structures (for example, culverts) and vehicles that use them.

The literature review and synthesis provides an overview of the current understanding, trends, and information gaps relating to the effects of roads and traffic on ecological conditions and the spatial scales at which roads affect ecological conditions. The available information at different scales of ecological effects is also examined.

## Approach

The committee collected and reviewed over 500 journal articles and conference proceedings. The literature was obtained primarily from scientific journals, although some reports were obtained from the grey literature. The list of studies was not meant to be exhaustive but nonetheless captured the majority of accessible literature. The focal area of the review was North America, but research findings from Europe and Australia were also included.

Table 3-2 summarizes the available bibliographic information. For each of the ecological conditions and spatial scales, the committee qualitatively categorized the number of studies as *none, few, several,* or *many.* For some subcategory and scale assessments, little information was obtained; however, there was a general consensus among committee members that more information was available but not in formats readily accessible. The findings of the studies are summarized in the following sections.

## Ecological Significance of Road Attributes

The presence or absence of roads is not the only factor that governs impacts on the surrounding environment. A major impact of roads is related to their use—that is, traffic. The density of the road network, the volume of traffic on a roadway or road segment, the road surface, and other engineered features also affect the extent of ecological effects of a road. This section briefly discusses the direct individual ecological impacts of various road attributes.

**TABLE 3-2** Summary of Number of Studies Addressing Different Types of Road Effects on Ecological Conditions

| Ecological Condition | Single-Segment Scale | Intermediate Scale | National or Regional Scale |
|---|---|---|---|
| **ABIOTIC** | | | |
| **Hydrology/Geomorphology** | | | |
| Stream networks | Few[a] | Many | None |
| Sediment production | Few[a] | Many | None |
| Changes in waterflow | Few[a] | Few[a] | None |
| **Chemical Characteristics** | | | |
| Mineral nutrients | Many | Few | None |
| Heavy metals | Many | None | None |
| Organic (water and sediment) | Few | None | None |
| De-icing salt | Many | Few | None |
| Volatile organic carbons (air) | Few[a] | None | None |
| **Other Disturbances** | | | |
| Landslides | Few[a] | Few | None |
| Light | Few | None | None |
| **BIOTIC** | | | |
| **Genetic** | | | |
| Structure | | | |
|   Barrier to movement | Few | Few | None |
| Functioning | | | |
|   Isolated populations | Few | None | None |
| Composition | | | |
|   Filtering effect | Few | None | None |
| Species/Populations | | | |
| Structure | | | |

| | | | |
|---|---|---|---|
| Wildlife population structure | Several | Few | None |
| Functioning | | | |
| Additional habitat | Several | Few | Few |
| Reduced habitat quality | Many | Few | Few |
| Dispersal corridor | Several | Several | None |
| Movement barrier | Several | Few | Few |
| Distribution | Several | Several | Few |
| Composition | | | |
| Species richness | Several | Several | None |
| Road mortality | Many | Several | Few |
| Nonnative plants in roadsides and adjacent landscapes | Many | Several | None[a] |
| Ecosystem | | | |
| Structure | | | |
| Fragmentation | Few | Few | None |
| Functioning | | | |
| Pollutants | Several | None | None |
| Composition | | | |
| Environmental characteristics | Several | Few | None |

[a]Categories thought to have more information available but not in readily accessible formats.

## Road and Traffic Density

The density of the road network, the volume of traffic on a road, the road's location, topography, and other factors have major roles in the intensity of associated environmental effects of roads. A few studies have correlated the density of the road network to their environmental effects (Findlay and Houlahan 1997, Carr and Fahrig 2001). Reduced construction of new roads reduces habitat fragmentation, suggesting that, in general, less habitat fragmentation occurs in a less-dense road network with high traffic volumes than in a dense network with low traffic volumes per road mile.

The direct effects of traffic density, defined in this report as the number of vehicle miles traveled on a given stretch of roadway in a given time, have been more widely studied. In general, an increase in traffic density correlates with an increase in the atmospheric deposition and the aquatic concentration of vehicle-emitted chemicals, such as heavy metals (Bocca et al. 2003, Fakayode and Olu-Owolabi 2003), particulate matter (Boudet et al. 2000), and organic pollutants (Ellis et al. 1997, Forman and Alexander 1998, Viskari et al. 2000, Ilgen et al. 2001, Latha and Badarinath 2003). Only areas with road density less than 0.72 km/km$^2$ (1.16 mi/mi$^2$) seem to support vibrant populations of wolves (*Canis lupus*) in Minnesota (Mech, et al. 1988, Fuller 1989), Wisconsin (Thiel 1985, Mladenoff et al. 1999), the western part of the Great Lakes region of the United States (Mladenoff et al. 1995), and Ontario (Canada) (Jensen et al. 1986). An exception to the trend is an established wolf population in a fragmented area of Minnesota with a road density of 1.42 km/km$^2$ (2.29 mi/mi$^2$) (Merrill 2000). Increased traffic density also has been shown to reduce amphibian population (Fahrig et al. 1995, Carr and Fahrig 2001).

## Road Surfaces

Because construction of asphalt concrete and hydraulic cement concrete road surfaces involves many of the same techniques, the differences in the direct effects of roadway construction using these materials are minimal. Well-constructed asphalt concrete and hydraulic cement concrete pavements are impervious; therefore, both are likely to exhibit similar runoff. Work completed in National Cooperative Highway Re-

search Program Project 25-09 and reported in 2001 in National Cooperation Highway Research Program (NCHRP) Report 448 (Nelson et al.
2001) concluded that most materials, including asphalt concrete and hydraulic cement concrete, used in the construction and repair of highways
"behave in a benign fashion in the environment. On the highway surface, leaching is slow, transport is rapid, and dilution is great...." Further, the results of Project 25-09 show no significant practical differences
in the potential impact of runoff from asphalt concrete and hydraulic cement concrete pavement surfaces.

## Engineering Structures

The impact of engineering structures is generally consistent with
the functioning of the structure itself. Concrete barriers, right-of-way
fences, noise barriers, and perhaps to a lesser extent, guardrails are designed to serve as barriers for people and noise, but they also function as
barriers to flora and fauna. These barriers may result in habitat fragmentation and species isolation (Forman and Alexander 1998).

Wildlife underpasses and overpasses, long-span bridges, and culverts can help to mitigate the adverse impacts of habitat fragmentation.
Examples of the use of these structures for ecological improvements are
identified in Chapter 4 of this report.

Poorly designed engineering structures can often hinder the
ecological improvements for which they were designed. In an aquatic
culvert system, for example, several key design characteristics ensure
effective utilization by the target species (see Box 3-1 under Biotic Consequences).

## Abiotic Consequences

Abiotic conditions that can be influenced by roads include hydrological, geomorphological, and chemical characteristics and such disturbances as landslides, noise, and light. In this section, the committee considers only changes to the abiotic conditions themselves, and examples
of each are provided below. How these abiotic changes affect the biota
is considered in later sections.

## Hydrological and Geomorphological Changes

Landscape changes result when roads alter the hydrological and geomorphological aspects of watersheds and landscapes. They can cause important changes (some for short periods, others for longer periods) in fluvial dynamics, sediment production, and chemical balances, which can adversely affect floodplain functioning and alter ecological conditions in aquatic and riparian areas (Figure 3-1).

Roads also affect water movements, sedimentation, and transport of pollutants. Because they often interrupt or otherwise alter sheet flow and

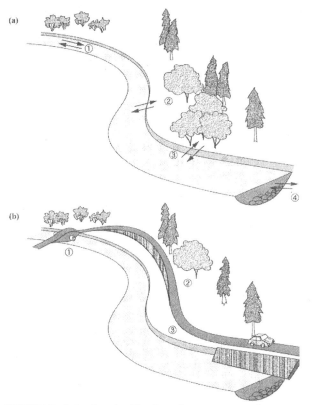

**FIGURE 3-1**   Road affecting four aspects of stream connectivity.   (a) Upstream-downstream (1), floodplain-stream (2), forest-stream (3), and surface-subsurface water connections (4).   (b) The connections severed or disrupted by a road in the floodplain.   Source: Forman et al. 2003.   Reprinted with permission; copyright 2003, Island Press.

runoff patterns, roads can affect the amount and quality of water that goes to recharging groundwater (Forman et al. 2003; NRC 1996, 2004), and they can affect surface waters in many ways. Because road embankments trap dust and dirt and they face the low winter sun at an angle, they can accelerate snowmelt (NRC 2003). Roads and associated ditches can become part of hydrological networks (Forman et al. 2003).

Several geomorphological processes and factors influence change. The nature of geomorphological processes affected by roads is strongly influenced by where and how roads are constructed, by the geology of the area, and by storm characteristics.

## Chemical Characteristics

*Water Quality*

The most observable abiotic environmental consequence of roads is the contribution of motor vehicles on paved roads to water pollution. However, this contribution cannot be disassociated from the surrounding land use.

The largest number of studies reporting on the chemical characteristics of road effects focus on the chemical effects arising from rainfall events at the single-segment scale (FHWA 1981; Asplund et al. 1982; Gjessing et al. 1984; Kerri et al. 1985; Lord 1985; Yousef et al. 1985; Barrett et al. 1995, 1998; Sansalone et al. 1995; Lopes and Dionne 1998; Wu et al. 1998). Water quality is adversely affected by pollutants present in surface runoff and the atmosphere. Pollutants that accumulate on roadways from spills, wastes generated during vehicle use, litter, and adjacent land uses enter waterways via surface runoff. Atmospheric wet (snow and rain) and dry (smoke and dust) deposition of pollutants, which can be transported long distances, also affects water quality and fisheries. Although concentrations of nitrogen oxide ($NO_X$) emissions from transportation have been quantified, there is no way to quantify wet deposition sources of nitrates from motor vehicles. Further, no quantification systems exist to measure or break down percentages of atmospheric nitrate deposition into a water body from specific nonmobile or mobile sources.

The primary source of pollutants associated with road use comes from vehicles, including fuel and exhaust; brake-lining and tire wear; leakage of oil, lubricants, and hydraulic fluids; and cargo spillage

(Forman et al. 2003, Hahn and Pfeifer 1994, Buzas and Somlyody 1997, Ball et al. 1998) (Figure 3-2). Shaheen (1975) examined urban roadway runoff and found that although the more hazardous constituents in highway runoff come directly from motor vehicles, they constitute less than 5% of the total solid pollutant load in highway runoff. These components include organic materials, such as petroleum and *n*-paraffin found in lubricants, antifreeze, and hydraulic fluids; lead; copper; chromium; zinc; nickel; and asbestos. Asbestos in brake linings was banned in 1989 (Shabecoff 1989), so vehicular sources of asbestos are minuscule, although resuspension of previously deposited asbestos is still a concern. In spite of the low contribution of constituents originating from the vehicle itself, vehicular traffic volume was identified as the principal factor influencing pollutant mass in highway runoff. That might be because vehicles are a transport mechanism as well as a source of pollution (Asplund et al. 1982).

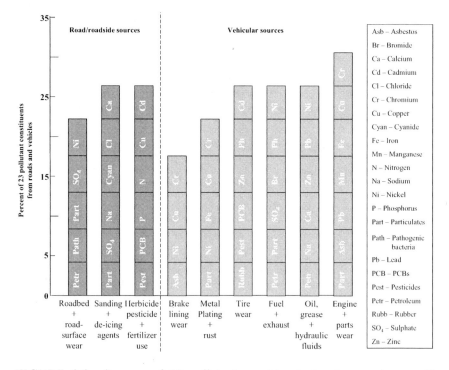

**FIGURE 3-2** Sources of 23 pollutant constituents in storm-water runoff. Source: Forman et al. 2003. Reprinted with permission; copyright 2003, Island Press.

More recent studies have associated chemical pollutants with traffic flow. Pollution associated with traffic varies not only with the density of the traffic but also with the ratio of passenger cars to trucks and the mechanical condition of the vehicles (Buzas and Somlyody 1997). Studies of the Brunette River Watershed in British Colombia showed a correlation between the concentration of hydrocarbons in streambed sediment and storm water and the density of traffic. The three predominant hydrocarbon types found—xylenes and alkyl-substituted benzenes, alkanes, and high-molecular-weight unresolved complex mixtures—are consistent with petroleum or petroleum-product sources, indicating a vehicle-related source. Values were between 2.6 and 3.4 milligrams (mg) of hydrocarbon per liter (L) of sediment or storm water per unit of traffic (vehicle km/day per hectare) (Larkin and Hall 1998).

There also is a strong correlation between concentrations of heavy metals and volatile matter in highway runoff (Flores-Rodriguez et al. 1994). Metals associate with organic matter, thereby changing the metal solubility, which primarily affects the temporal characteristics of the runoff (Harrison and Wilson 1985).

Road salt has been commonly used to de-ice roads for many years. Compared with the literature on other road-related contaminants, the literature on pollution of surface water and groundwater by road salt is voluminous (Forman et al. 2003). The use of road salt results in the accumulation of sodium and chloride ions in runoff, thereby increasing concentrations of those ions in the soil, groundwater, and surface water above background concentrations and sometimes to unacceptable concentrations in drinking-water sources. The increase in the concentrations of ions reduce the soil's ability for ion exchange, decreasing permeability and aeration, and increasing alkalinity of the soil.

## Air Quality

Some studies also focused on the impact of vehicular chemical pollutants on local air quality. The majority of these studies examined the impact of vehicular traffic on the presence or absence of volatile organic compounds (VOCs) (Clifford et al. 1997, Tsai et al. 2002). Surprisingly few studies have examined the effects of chemical pollutants at the intermediate scale that could provide valuable information on total area effects primarily in watersheds or protected areas. Although most of the concern with roads and air quality focuses on new emissions added by vehicle tail-pipe emissions, resuspended particles from traffic flow, dust from roadside areas, and other fugitive (non-tail-pipe) emissions are of concern.

The primary effects of local air pollution come from the increase in VOCs, $NO_X$, carbon dioxide, particulate matter, and ground-level ozone that come from emissions of the traffic on the road. In the presence of sunlight, VOCs and $NO_X$ are precursors of ozone, a regulated ambient air constituent in the United States (de-Nevers 2000). The impact of these pollutants on a variety of species, primarily humans, has been well documented (Forman and Alexander 1998, Ilgen et al. 2001, Buckeridge et al. 2002, Delfino et al. 2003, Dongarra et al. 2003, Wilhelm and Ritz 2003, Zmirou et al. 2004).


## Other Disturbances

### Landslides

Physical disturbance can disrupt ecological systems, and roads promote such disturbances. For example, roads in mountainous areas can create landslides due to unstable soil and steep slopes. Paved road surfaces can increase water discharge rates in watersheds, thus increasing the potential for landslides and flash floods in streams and rivers.


### Lighting

Roads and associated structures usually have artificial lighting. At some interchanges, especially near urban centers, the lights can be intense. Many rural roads do not have lights, although headlights from nighttime traffic and occasionally other lights are visible.


### Noise

Noise along roads is a function of traffic type and amount. In rural areas, road noise can be audible to humans up to 10 km from the road and occasionally more than 10 km in optimal conditions.

*Local Climate Effects*

Roads interact with climate at a wide range of scales. At local scales, highly developed areas (urban centers) have been shown to experience an increase in temperature (Woolum 1964) in a process called the urban-heat-island effect. Urban heating can also result in increased rainfall (Shepherd et al. 2002). Roads change the albedo (fraction of light reflected by a surface) and other surface characteristics, but other structures, such as buildings, parking lots, and sidewalks, also contribute to heat-island effects.

Local climate might also be affected simply by the presence of roads and associated development. The loss of pervious surfaces and vegetation and their replacement with impervious surfaces that hold heat and do not respire result in localized temperature increases. Temperature increases can result in increased volatilization of organic contaminants from vehicular emissions (Saitoh et al. 1996). Thermal characteristics of the road surface cause accelerated snow melt (NRC 2003).

## Biotic Consequences

Roads can have biotic effects on the genetics of populations, on species, and on ecosystems, and their effects can accumulate over space and time (e.g., NRC 2003; Figure 3-3).

The framework prepared by the Environmental Protection Agency (EPA) (Figure 6-1) also is a helpful way to conceptualize the ecological effects that roads can have. In general, their effects can operate through a variety of ecological mechanisms.

The effects occur at various stages of road planning, construction, operation, maintenance, and perhaps decommissioning or road removal. They often are expressed differently over space and time. Any approach to the assessment of the ecological effects of roads must take these broad categories of effects into account, as well as the variety of scales over which they can operate. Below, the various categories of effects are discussed. The biotic consequences of the following effects of roads are considered below: *direct effects* include roads as barriers, enhancement of dispersal, roadkill, and effects on habitats; *indirect effects* include results of the access that roads provide to previously inaccessible areas, changes in water and air quality, and effects of lighting and noise.

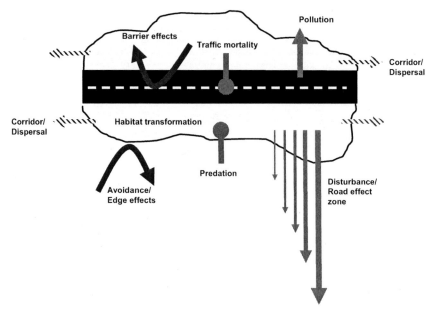

**FIGURE 3-3**  Schematic representation of the primary ecological effects of roads on species and populations.  Source:  Adapted from van der Zande et al. 1980.  Reprinted with permission; copyright 1980, Elsevier.

## Roads as Barriers

Roads can impede animal movements by direct mortality or avoidance behavior.  The barrier effect varies between species, road types, and adjacent habitat quality; however, traffic volume and speed strongly influence the effect.  Some authors have suggested that divided highways with 90 m of cleared areas as barriers are as effective as bodies of water twice as wide in obstructing dispersal of small forest mammals (Werner 1956, Sheppe 1965).  In the Canadian Rockies, grizzly bears were more likely to cross low-volume roads and more likely to cross at points with high habitat rankings.  Male grizzly bears were found closer to low-volume roads than females, but they crossed roads less often than females, particularly during the berry season (Chruszcz et al. 2003).

The barrier effect for some species is less related to traffic than to habitat changes (road-forest edges and gap creation caused by roads).  Small road clearances (less than 5 m) can impede movement of certain small mammals.  For example, road crossing by small mammals was inversely related to road width in Australia (Barnett et al. 1978), and small road clearances (less than 3 m) have been shown to reduce crossing by small mammals, such as voles and rats (Swihart and Slade 1984).

Barriers to the movement of wildlife can lead to fragmentation of populations. Isolation caused by physical barriers to movement, such as roads, may reduce gene flow, thus causing genetic effects (Slatkin 1987) that in the extreme could result in local extirpation. For small mammals, that could result in ecosystem-level alterations because of their importance as seed dispersers and their role as prey for such predators as marten, wolverine, and raptors.

Fish passage can be blocked by improperly functioning stream culverts (Box 3-1) or by a lack of them, creating an often impassable barrier. The committee is not aware of studies showing that culverts have genetic effects on aquatic organisms, although such effects could be expected. Knaepkens et al. (2004) used genetic analyses to show that a culvert was not a migration barrier for the endangered European bullhead (*Cottus gobio*), but Schaefer et al. (2003) reported that culverts did restrict movement of the darter (*Percina pantherina*, a North American fish).

Because little is known about the long-lasting ecological effects of roads on animal populations, concern has been raised about the large-scale influence of barriers, such as interstate highways, on normal mammalian distributional patterns and perhaps ultimately on speciation (Baker 1998). Mitigation measures, such as large-span bridges and wildlife-crossing structures, are successfully being used to reconnect isolated populations, restore hydrological processes, and assist movement of wildlife across roads (Forman et al. 2003) (Figure 3-4, see also Chapter 4).

By creating barriers, roads also affect ecosystem functioning. The effects of roads on hydrological processes, most commonly through interfering with patterns of flow, have been the focus of many studies. Changes in hydrological processes affect ecosystem processes, such as habitat connectivity, primary productivity, decomposition, nutrient cycling, and disturbance regimes (for example, flooding frequency and intensity) (Jones et al. 2000).

## Roads as Enhancers of Dispersal

Roads can act as habitat corridors as well as barriers. Road corridors (roads, their verges, and sometimes roadside ditches or wetlands) can contribute to the movement of wildlife and plant dispersal. Roadside verges facilitate animal movement, resulting in range expansion or dispersal between core habitats. Verges also aid in the spread of plant spe-

**BOX 3-1** Aquatic Culvert Design Effects

Salmon have evolved to negotiate waterfalls during both up-stream spawning migrations and downstream juvenile passage to the sea. However, this ability is fairly rare among fishes. Few other families can accomplish either, especially with respect to upstream movement. Hence, even a small vertical rise or drop becomes impassable. The general image of a dam may be a wall of rock or concrete, but in fact any steep drop in flow of more than a few centimeters is impassable to most fishes.

Among the most common such barriers are road culverts, typically involving large pipes under a road where it crosses a stream. If a culvert is not level with the grade of the stream, if the stream gradient is more than a few degrees, or if velocity exceeds 1 m/sec (and even if a culvert initially lies level with the bed of the stream), flow around and through the pipe scours and erodes the streambed. Hydraulic jump upstream and outlet drop downstream create an ever-steeper waterfall and deeper plunge pool. The pipe mouth eventually extends out over the stream, creating a vertical drop while lacking the rock face characteristic of a natural waterfall.

Culverts and similar road-crossing structures have proved to be substantial barriers to fish passage. In a study of spring and summer movement by 21 fish species in seven families in the Ouachita Mountains of Arkansas, Warren and Pardew (1998) found an order of magnitude less movement upstream through culverts than through other types of crossings or natural reaches. They also found that fishes upstream of culverts were significantly less likely than fishes below culverts to move downstream, a result they attributed to avoidance of the increased flow velocity that typifies culverts. Such reluctance to move could be a factor isolating upstream populations and contributing to localized extirpations. Culverts can increase vulnerability of imperiled species by reducing movement among habitat patches. For example, federally threatened leopard darters (Percina panthera) in Oklahoma failed to move upstream through culverts, even though water temperatures at downstream sites had risen to undesirable levels and thermal refugia were available upstream (Schaefer et al. 2003).

Obstructions, such as culverts, and their impacts are avoidable. Designs for culvert construction that minimize impacts on fishes are readily available (TranSafety 1997, Moore et al. 1999, Bates et al. 2003) and take into consideration hydrological, geological, biological, and economic factors. Where conditions prohibit design modifications that minimize culvert impacts, small bridges become a preferred, albeit more expensive, alternative.

**FIGURE 3-4** Wildlife crossings are designed to link critical habitats and provide safe movement of animals across busy roads. Typically they are combined with high fencing and together are proven measures to reduce roadkill and restore movements and regional connectivity. The photograph is of a newly constructed open-span bridge underpass installed on the Trans-Canada Highway near Canmore, Alberta. Source: Photograph by Tony Clevenger, 2005.

cies (often exotics). Roads with low traffic volumes are often used by wide-ranging wildlife because of the ease of travel, particularly when snow is present.

Invasion of nonnative plants can also occur from vehicles transporting nonnative seeds into natural areas and clearing land during road construction (Tyser and Worley 1992, Parendes and Jones 2000, Gelbard and Belnap 2003). In addition, insects and pathogens can be transported into new environments by vehicles (NRC 2002).

## Roadkill

Roadkill can have demographic consequences for some species of wildlife (Maehr et al. 1991, Jones 2000). Roads and traffic can reduce wildlife population densities and ultimately affect the survival probability of local populations. Traffic-related mortality has contributed to the decline of several species: Eurasian badger (*Meles meles*) (Bekker and Canters 1997) and moor frog (*Rana arvalis*) (Vos and Chardon 1998) in The Netherlands, Hermann's tortoise (*Testudo hermannii*) (Guyot and Clobert 1997) in southern France, and Florida panther (*Felis concolor coryi*) (Maehr et al. 1991) are some examples. Road networks also particularly affect wide-ranging carnivore species (Maehr et al. 1991, Brandenburg 1996). Metapopulation theory suggests that more mobile species are better able to manage with habitat loss (Hanski 1999). Yet mortality of individuals in the matrix habitat (for example, road corridors) does not typically figure into metapopulation theory. Studies show that when mortality is high in the matrix habitat, highly mobile species are actually more vulnerable to habitat loss (Carr and Fahrig 2001, Gibbs and Shriver 2002). Younger age classes tend to be more affected by roads, as they interact more and live closer to them (Fowle 1996).

## Habitat Effects

Roads have large, widespread effects on aquatic habitats (NRC 1996, 2004; Forman et al. 2003). When roads fail, landslides and torrents of water-borne debris can have serious adverse effects on stream habitats (NRC 1996). Roads and their associated structures, such as bridges, culverts, and berms, modify streamflows and sediment transport and often make passage for aquatic organisms more difficult or even impossible (NRC 1996, 2004; Forman et al. 2003; Warren and Pardew 1998; Schaefer et al. 2003). Because paved roads (and to a lesser degree, unpaved roads) are impervious, they increase runoff and otherwise alter hydrological patterns. Finally, they often interrupt the connectivity of aquatic ecosystems, although by providing new networks of aquatic systems, for example, in long ditches, they can enhance connectivity as well (Forman et al. 2003). Fragmentation effects of roads, as part of the cumulative effects of many factors, can strongly influence the distribution and land-use patterns of wide-ranging and migratory wildlife (Ward 1982, Noss et al. 1996).

Roads affect types of landcover, particularly their spatial composition and structural integrity. Although roads are narrow and linear, as discussed previously in the chapter, they create a disproportionate amount of habitat fragmentation, resulting in loss of connectivity. Although other factors contribute to fragmentation, roads are clearly a major factor. There are various ways to measure habitat fragmentation that reflect different ecological concerns, such as number of patches, patch size, change in patch size, number of edges, edge size, and the nature of the barrier. As road density increases, larger and more contiguous expanses of habitat become smaller and more isolated (Forman et al. 2003). Substantial amounts of edge habitat with increased heterogeneity are created by roads and benefit "edge" species, such as white-tailed deer, whereas interior "core-sensitive" species are at a disadvantage (Forman et al. 2003).

Although roads typically adversely affect the density and diversity of plant and animal species and populations, especially early on, there are locations along roads and bridges over waterways that support healthy and diverse wildlife populations. For example, in Virginia at the crossings of the Potomac and Rappahannock Rivers along the I-95 corridor, there are populations of eagles and osprey that feed on the fish near the road-bridge crossings. Although these particular species are supported in this area, it might not be their preferred habitat, and the density and diversity of other wildlife in this area might be different from that during preconstruction.

Roadside verges (margins) can increase habitat diversity where there is little remaining natural or seminatural habitat. Depending on the nature of roadside verges, they can support abundant populations of some small mammals, insects, and birds, as well as native plant species. Roadside verges can also be important habitats for rare native plant species when juxtaposed in human-modified landscapes. For example, a large proportion of native plants and animals in Great Britain are found in roadside verge habitat. Furthermore, roadside verges have the potential to help restore native grass and wildlife communities. Also, in the arid west, the productivity of desert vegetation can be markedly greater near the roadside (Johnson et al. 1975), particularly on the upslope side. However, large areas on the downslope side of the road can be deprived of runoff, resulting in much lower plant biomass and productivity (Schlesinger and Jones 1984).

Roads create ideal conditions for the proliferation of nonnative plants because of increased light, disturbed soils, and dispersal vectors (caused by wind and vehicles). Many nonnative plants were introduced intentionally to control erosion; however, there are now ongoing efforts to reduce the use and spread of nonnative plants on roadsides and in adjacent habitats. The practical aspects of nonnative plant management and other resource issues are discussed further in Chapter 4.

## Road Access with Secondary Effects

Roads are usually built to provide transportation corridors between population centers or between industrial or resource centers and users. In such cases, development and resource use, or the expectation of them, precede and motivate the construction of roads. It can be difficult to separate cause and effect in many cases, but an excellent example of a road built to access planned development is the Dulles Airport Access Road in northern Virginia, which was specifically built to provide that access and which prohibits use by non-airport traffic.

However, roads also can bring about development or environmental exploitation by providing access to secondary resources (not the resources that the road was built to access). The provision of secondary access often also involves the construction of secondary roads (e.g., Hawbaker and Radeloff 2004). The construction of the James Dalton Highway from Fairbanks to Prudhoe Bay on Alaska's North Slope to service the Trans-Alaska Pipeline and provide road access to the large oil fields also has provided access to the North Slope for recreational hunters, anglers, and tourists, increasing the mortality of some animal species (NRC 2003). Forest roads in the Congo Basin (Africa) have had the secondary effects of providing market accessibility and access to animals by hunters (Wilkie et al. 2000), and roads can influence patterns of settlement and land use (e.g., Pedlowski et al. 1997, Wear and Bolstad 1998, Hawbaker and Radeloff 2004). A major bypass being constructed around the Hoover Dam on the northwestern Arizona and southeastern Nevada border has caused developers to plan to build 55,000 houses in sparsely populated White Hills, Arizona, because the new road will cut the driving time to Las Vegas, Nevada, by at least one-third, making commuting from there feasible (Argetsinger 2005).

## Water Quality

Because roads are impervious, runoff from them is greater than that from most unconstructed land cover types. As mentioned earlier, this runoff usually contains pollutants from the road and the vehicles on it.

Wegner and Yaggi (2001) reviewed the environmental impacts of road salt and of its main alternatives, calcium magnesium acetate and potassium acetate. They reported a variety of adverse impacts of salt on roadside and aquatic plants and animals. Ferrocyanide used as an anticaking agent in road mixtures can also harm sensitive fishes (Environment Canada 2003).

Runoff contaminated with chemicals, including salt, affects roadside vegetation. Salt runoff can damage vegetation, resulting in reductions in seeding establishment and flowering and fruiting of sensitive plant species; foliar, shoot, and root injury; and growth reductions. Areas affected by salt runoff also demonstrate a shift in plant community structure when salt-sensitive plant species are replaced by halophytic species, such as cattails and common reed grass (Environment Canada 2003). Fleck et al. (1988) reported that roadside salt and sand application resulted in significant declines in the vigor of roadside stands of white birch (*Betula papyfera*), including increased numbers of dead trees. Richburg et al. (2001) reported that both salt and invasive species affected the species composition in a roadside wetland and suggested that the presence of roads could have enhanced the ability of the invasive giant reed (*Phragmites australis*) to invade the ecosystem.

The effects of road salts on vegetation discussed above can affect wildlife in several ways, including by diminishing their habitat (Wegner and Yaggi 2001, Environment Canada 2003). Road salting can inhibit movement across roads by amphibian species. Survivorship of some amphibians was also reduced in roadside pools contaminated by road salt. Behavioral and toxicological impacts have also been associated with exposure of mammals and birds to road salts. Ingestion of road salts increases the vulnerability of birds to vehicle collisions and may poison some birds if water is not available.

$NO_X$ are emitted by vehicles and are deposited from the atmosphere either through smoke and dust (dry deposition) or through rain and snow (wet deposition). Both kinds of deposition contribute nitrogen to aquatic environments, such as the Chesapeake Bay. Their effect accumulates with the effects of other sources of nitrogen in runoff, mainly agriculture, and they can result in algae blooms. Algae reduce the penetration of light, thus killing submerged aquatic vegetation that forms the basis of the bay's food chain. As the algae die and decay, oxygen levels decrease, further affecting aquatic organisms. A diverse range of effects can be generated by many sources of chemical pollutants, including sediment, oil and grease, metals, and organics. Pollutants in runoff from highways are comparable to those of urban runoff (Ellis et al. 1997). The organic chemicals tend to adhere and partition to particulate matter and to accumulate in aquatic organisms. Many of these organic pollutants are known or suspected to be carcinogens and are listed as priority pollutants by the U.S. Environmental Protection Agency.

## Air Quality

Alterations in roadside plant communities can also change chemical characteristics of an ecological system. Chemical pollution from vehicle exhaust (primarily $NO_X$) enriches roadside soil and changes plant composition, favoring a few dominant flowering plants at the expense of more sensitive plant species (for example, ferns, mosses, and lichens). The extent of this effect can range up to 200 m from multilane highways and up to 35 m from two-lane highways (see previous section for discussion on the effects of nitrogen dissolved in water).

## Lighting

Artificial lighting associated with dams and canals in The Netherlands did not affect the spatial behavior of most mammal species; however, predators (such as stoats, foxes, and polecats) appeared to be attracted to the lights (de Molenaar et al. 2003). The study did not evaluate any effects of lighting, noise, and movement associated with the presence of roads and traffic. Other studies examining lighting pollution in general, however, suggest that roadway lighting has ecological effects (Scigliano 2003). Beier (1995) found that cougars avoided night lighting in a fragmented and increasingly urbanized habitat in southern California. Lighting in the study area was associated with industry and roads, including residential and busy highways. More research is needed on the effects of artificial lighting on animal movements to better understand and properly assess how lit roadways and increasingly lit landscapes affect the long-term survival of wildlife populations.

## Noise

Impacts of noise from vehicle movement on humans have been well documented, but noise also affects wildlife (Forman et al. 2003). For example, of 43 species of woodland breeding birds, 26 species (60%) showed reduced densities near highways (Reijnen and Foppen 1994). Traffic noise explained the most variation in bird density in relation to roads in a regression model. This effect also occurred for grassland birds (Reijnen et al. 1996) and is more important in years with a low overall population size (Reijnen and Foppen 1995). An analysis of the total effect of The Netherlands's most dense network of main roads on "meadow birds" showed a possible population decrease of 16%, attributable to reduced habitat quality and traffic noise (Reijnen et al. 1997).

It is not known how well wildlife can acclimate to constant noise, for example, along a roadside, and acclimation no doubt varies among organisms. Furthermore, breeding activities of species, such as birds and amphibians that rely on vocalization, may be particularly susceptible to disruption by noisy conditions. Existing noise barriers in suburban settings are designed to protect humans from noise. As such, they can serve as barriers to animal movement. Generally, noise impacts decline with distance to the road.

## Range of Occurrence of Effects

Roads interact with plants, animals, water, sediment, and other ecological attributes in ways that extend beyond the road edge (Table 3-3). The distance from a road that ecological effects can be detected is called the "road-effect zone" (Forman et al. 2003) or "zone of influence" (NRC 2003). The effect of distance varies, depending on the organism, location, and disturbance type, and generally increases with traffic volume (Clark and Karr 1979, Reijnen and Foppen 1994, Nellemann et al. 2001). Aquatic environments and organisms are highly sensitive to roads and traffic densities (Eaglin and Hubert 1993, Vos and Chardon 1998, Turtle 2000). For example, wetland species diversity is negatively correlated with paved-road density up to 2 km from wetlands (Findlay and Houlahan 1997). The effects of roads on wetland diversity take about 3-4 decades to be fully realized (Findlay and Bourdages 2000).

Road-effect zones occur due to disturbances from high-volume traffic, which can reduce the habitat quality near roads (Table 3-3). Breeding densities and distribution of many bird species are reduced adjacent to busy roads. Animals avoid roads by a distance that increases with increasing traffic volumes. This road-avoidance zone contributes to the road-effect zone. Similar distance effects of roads occur with chemical

**TABLE 3-3** Examples of the Extent to Which Road-Induced Effects Penetrate Adjacent Habitat

| Road Effect | Distance | Reference |
|---|---|---|
| Heavy metals | | |
|   In soils and plants near roads | 50-100 m | Ministry of Transport, Netherlands 1994 |
| Chemical pollution | | |
|   Oxides of nitrogen changing plant communities | 200 m | Angold 1997 |
| Animal distribution | | |
|   Pink-footed geese (*Anser brachyrhunchus*) and graylag | 100 m | Keller 1991 |
|   geese (*A. anser*) | 150 m | Ortega and Capen 1999 |
|   Territory size of ovenbirds (*Seiurus aurocapillus*) | | |
| Traffic noise | 200-1,200 m | Van der Zande et al. 1980 |
|   Breeding bird density | 40-1,500 m | Reijnen et al. 1996 |
| Road lights | | |
|   Breeding bird density | 200-250 m | De Molenaar et al. 2000 |
| Avoidance zone | | |
|   Caribou (*Rangifer tarandus*) | 5,000 m 200 m | Nellemann and Cameron 1998 |
|   Deer (*Odocoileus hemionus*) and elk (Cervus *elaphus*) | 1,000 m | Rost and Bailey 1979 |
|   Grizzly bears (*Ursus arctos*) and black bears (*U. americanus*) | | Kasworm and Manley 1990 |
| Increase in edge species | | |
|   Component of bird community | 100 m | Ferris 1979 |
| Road density | | |
|   Wetlands species richness | 2,000 m | Findlay and Houlahan |
|   Moor frog (*Rana arvalis*) presence | 750 m | 1997 |
|   Leopard frog (*R. pipiens*) distribution | 1,500 m | Vos and Chardon 1998, Carr and Fahrig 2001 |
| Early melting of permafrost | 100 m | Walker et al. 1987 |

pollution, nonnative plant species, and other wildlife species' distributions. The road-effect zone is reduced on low-volume roads.

## Scale of Effects

Most of the literature reviewed by the committee focused on the effects of roads on species and populations of wildlife and plants. The need to assess project-level effects of road building and expansion on species and their populations as part of policy guidelines (the National Environmental Policy Act and the Endangered Species Act) has been the catalyst for most single-segment and intermediate-scale studies (Evink 2002). Although current research is making valuable contributions, its ultimate impact is limited by low funding, inadequate coordination across research entities, and short-term or project-specific focus (TRB 2002a).

Species and populations have been relatively well researched at the intermediate scale, particularly with respect to the effect of roads on species composition. Little information has been reported at the national scale on species richness, nonnative plants at roadsides, and road-related mortality of wildlife. These effects, although varied, are widespread geographically. Thus, a synthesis of available information or meta-analysis could provide valuable insight into the extent of effects at a broader scale than is known today. The committee found that very little research covers long periods, and almost no research has addressed large spatial scales of road effects.

Few studies were found that describe the effects of roads on ecosystems, and most of those were carried out at the single-segment scale. Investigations of the effects at larger scales are scarce. A national-scale assessment estimated that one-fifth of the U.S. land area is directly affected ecologically by public roads (Forman 2000), even though the paved road network covers less than 1% of the U.S. land area. With increasing availability of digital biophysical and land-use data, geographic information system (GIS) tools and applications are becoming widely used among resource managers and transportation planners for amassing information and modeling the potential effects of roads at multiple scales (Dale et al. 1994, Tinker et al. 1998, Vos and Chardon 1998, Clevenger et al. 2002a). The increased use of GIS will probably facilitate more GIS-based studies that evaluate the ecological effects of roads at regional and national scales.

Ecologists have long conducted studies on species diversity patterns at broad spatial and temporal scales (Brown and Lomolino 1998). Yet, attempts to understand how geographical and environmental features structure genetic variation at the population and the individual levels are new (Manel et al. 2003). These approaches focus on processes at fine spatial and temporal scales by detecting genetic discontinuities and correlating these with such environmental features as barriers, including highways (Gerlach and Musolf 2000, Conrey and Mills 2001, Thompson 2003). The new genetic approaches and techniques combined with in creasing interest in how highways affect population viability is likely to result in more research in the coming years.

The committee selected several of the most common ecological effects of roads and plotted the extent of their effects with respect to spatial and temporal scales (Figure 3-5). The abiotic consequences of altered water flow and sediment deposition are relatively fast-acting and essentially limited to single-segment and watershed scales. However, the impact of chemical pollutants, both organics and heavy metals, can be long lived (such as contaminated drinking-water sources) and far reaching (such as atmospheric deposition) due to sediment and particulate transport, sediment accumulation, and bioaccumulation. For the biotic component, reduced genetic structure related to barrier effects on animal movement probably would be manifested over a longer period (months and decades) and have effects at a broader spatial scale (watershed and eco-regions) than most abiotic effects. Proliferation of invasive exotic plants and landscape fragmentation due to roads occur over longer periods and are pervasive, affecting entire nations and continents as well as smaller-scale areas.

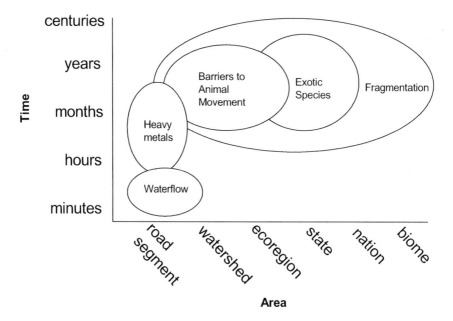

**FIGURE 3-5** Spatial and temporal dimensions of ecological effects of roads.

Figure 3-5 reveals some of the spatial and temporal scales at which these effects most often occur. (See Figure 6-4 for another plot of spatial scales by the U.S. Department of Defense.) The figure suggests that ecological effects of roads can alter ecological processes over scales that range from minutes to centuries (time) and from meters to thousands of kilometers (in space). That is, ecological effects occur at scales that extend much longer in time and broader in space than those scales currently being used in assessment, planning or management. This issue is addressed in subsequent chapters. The figure also suggests that there is no "correct" scale for understanding how roads interact with ecosystems, but rather multiple, overlapping ranges of scales that correspond to specific ecological structures and processes under consideration.

Figure 3-5 also indicates that ecological processes occur at particular scale ranges, and hence assessments conducted at different scales could miss some ecological effects of roads. An example was discussed in a recent National Research Council report (NRC 2003). In that case, (unpaved) roads and the traffic on them affected the movements of caribou, especially of females. As a result, the caribou were more likely to encounter insects in years favorable to insects, and the interaction between roads and insects resulted in a subtle but measurable reduction in carbon productivity in those years. Local assessments did in fact fail to identify that effect; the assessment that was required extended over a far greater area than the direct road-effect zone for caribou. Other cases might include birds that migrate between wintering and breeding areas over many thousands of kilometers. A road that affects a resting area on a migration route might affect the entire population of the species, but conducting the assessment at a scale of hundreds of meters—or even at the scale of a county—could easily miss that population effect. A very similar example would be a road that crossed a stream that provided habitat for a migratory species of fish (see for example Box 3-1 and the entire discussion in the section "Roads as Barriers" in this chapter). An assessment at the scale of the barrier would most likely miss any population effect.

## Cross-Scale Effects

This section indicates that ecosystems are scale variant; that is, the cross-scale biological structures and processes cannot be easily aggregated from one scale to another but are dependent on the scale of focus.

Scale variance is in contrast to many physical processes, such as water-flow dynamics and large-scale fire patterns that are scale invariant (Gunderson and Snyder 1994). Scale invariance means that structures and processes are self-similar as the scales change. A broad set of human-induced ecological changes, such as forest-pest outbreaks, algae blooms, salinization, and grassland to shrubland conversion, are all examples of scale-variant phenomena. The property of scale variance has great implications for the ability to assess and manage across a wide range of scales.

The major conceptual framework for understanding ecological cross-scale structure and dynamics is hierarchy theory. Ecologists (Allen and Starr 1982, O'Neill et al. 1986, Allen and Hoekstra 1992) built on the seminal work of Simon (1962) to develop a theoretical base that emphasizes a pattern of aggregations (hierarchical levels, or "holons") nearly separable across scales. Hierarchical levels can be identified by a stronger set of interactions within a hierarchical level than among levels. These hierarchical levels correlate to scales, and each level has characteristic spatial and temporal domains; that is, each level (see leaf, tree, or forest level in Figure 1-9) has a characteristic turnover time and spatial domain. Hierarchical levels can be identified in ecological systems, but how do they interact?

The nature of ecological interactions at various scales has been the subject of much scientific debate. The focus has been on the interaction between processes and their associated structures that operate for long periods and over large spatial scales and processes that are faster and smaller. They also have been cast as "top-down" versus "bottom-up" control. An example of top-down control, sometimes called "hierarchical control," is how altered soil types and microclimates associated with road rights-of-way determine the suite of plant and animal species that thrive. Because roads affect variables that change slowly, such as geological formations, soil composition, and topography, they often produce top-down effects. However, top-down and bottom-up effects often occur together and interact; this complex and dynamic set of ecological interactions has been called "panarchy" by Gunderson and Holling (2002). Examples of such interactions include disturbance dynamics, such as forest fires or forest-pest outbreaks (Gunderson and Holling 2002). Peterson (2002) demonstrated how a road network can disrupt spatial patterns and succession in southern U.S. forests. Other types of surprising ecological behavior appear to arise from such panarchical interactions.

## Scales of the U.S. Road System and Ecological Effects

The physical structure of the U.S. road system covers few thousands of kilometers. The replacement time of roads is on the order of multiple decades, although this value can vary as a function of road type. The presence of roads across the landscape generates a variety of effects and interactions with ecological systems. The effects fall into three categories: (1) effects that are fixed in scale; that is, the domain of the change is fixed in space and time with sharp boundaries; (2) effects that generate or initiate cross-scale interactions; and (3) effects that constrain or limit cross-scale interactions. In this context, cross-scale interactions are those that traverse hierarchical levels. Each of the categories is described in the following paragraphs.

Many ecological effects of roads are spatially small. Most of the documented effects occur at the road-segment level, which includes the road, roadside, and a zone described as the effective road-impact zone (Forman et al. 2003). Generally, the zone ranges from a few meters to a few kilometers, depending upon the type of impact. Many effects are confined to the road and shoulder zone. Altered physical and chemical soil conditions from construction, management (fertilization or salt applications), or vehicle exhausts are found primarily in a narrow zone around roads. Some of the effects on biota, such as changes to populations (increased mortality) or community composition, occur primarily within this zone or within an area of a few hundred meters perpendicular to the road segment.

Some road effects cross scales of space and time. These effects include the abiotic and the biotic components of ecological systems. One example is the set of effects on hydrological systems, where sediments, nutrients, and heavy metals are introduced into riparian systems. Changes in inputs have created shifts in biogeochemical cycles, resulting in changes in species distribution and abundance. Another cross-scale effect occurs when roads serve as ecological corridors and increase dispersion. One example attributed to roads is the spread of exotic organisms, both plants, such as kudzu, and animals, such as armadillos (Taulman and Robbins 1996).

Many ecological effects of roads eventually influence structures and processes at longer and broader scales than first imagined. The spread of kudzu (or any other invasive organism) is a good example of an unintended, broader scale effect. Originally intended as an ornamental vegetation cover for stabilizing steep roadsides, the vine has spread across much of the southeast, invading areas never imagined in original assessments. Similar arguments could be made for the nutrients, such as nitrogen from automobile exhausts, that have been observed to spread via atmospheric transport and accumulate in wetland and estuarine areas.

Cumulative effects have been described in two ways. One type of cumulative effect is the cross-scale effect described in the previous paragraph. These effects accumulate over time, space, or both. They can manifest as a cumulative change in an ecosystem structure or function, such as the increase in the amount of heavy metals in soils adjacent to roadways or the increase in species or populations, such as scavengers that eat organisms killed on the roadway. Spatial accretions occur as structures increase in distribution, such as the spread of kudzu. The other type of cumulative effect involves synergistic interaction among key structures or functions associated with a road. For example, caribou migration in Alaska was differentially affected by the combination of roads and oil pipelines (NRC 2003). The implications of the latter type of effect for assessment and environmental review are discussed in the next section on cumulative effects.

The final set of effects occurs when roads decrease the scale of ecological structures or processes. Often, they are barriers to landscape-scale phenomena. The restriction of wildlife migration or dispersion has long been recognized as a road effect. Fragmentation of populations due to roads is another such effect. Broad-scale disturbances, such as fire, that are critical to many types of ecosystems (prairies in the Midwest and pine forests in the eastern and western United States) are limited in spatial extent by roads that act as fire breaks. In some fire-adapted ecosystems where fire is heavily managed, rules of smoke management restrict when and how prescribed fires may be set because of the need to prevent visual hazards on roads caused by the smoke. Shifts in the timing and extent of fires can generate large-scale changes in ecosystems.

## Cumulative Effects

Even though the awareness of the ecological effects of roads has grown steadily over the past few decades, only a small body of literature addresses cumulative effects associated with roads. In evaluating effects of oil and gas activity on Alaska's North Slope, the NRC (2003) found that roads had a synergistic effect with pipelines and off-road vehicle traffic. These factors accumulated to affect the habitat and behavior of animals, physically changed the environment next to the road, and increased access and social contacts among human communities. Caribou migration on and near roads, although they are gravel, is one example of a cumulative effect; roads with a parallel pipeline and those without a parallel pipeline had different migration patterns (Forman et al. 2003, NRC 2003). In addition, new roads often are associated with development of residential, commercial, and industrial activities. In some cases, the roads are built to support the new activities and, in other cases, the roads lead to the additional development.

## INFORMATION GAPS

Historically, most studies of road effects have been carried out at the project level, with local studies focusing on specific transportation effects. Collaborative research among multiple government agencies has been lacking. States or provinces have had little coordinated formal data sharing to allow for information syntheses and analyses of effects at larger and perhaps more meaningful scales of evaluation. Defining the appropriate scale of research will depend on the ecological condition of interest. A watershed is one example of an appropriate spatial scale to assess water-quality issues. Transportation projects sponsored by the Federal Highway Administration (FHWA) and the Transportation Research Board (TRB) have stimulated and encouraged collaborative studies involving multiple state agencies with similar transportation problems that might be solved through large-scale ecological assessments and pooled-funding approaches (TPF 2004). Pooled-funding projects often focus on specific information needs defined by the collaborative state transportation agencies, which frequently represent diverse environments across the continent. Often, projects are not defined by appropriate scales or ecologically defined areas, such as specific eco-regions (for example, the northern Rocky Mountains).

Reports have called for nationwide assessments and national syntheses on how wildlife respond to highway barriers, for mapping habitat linkages and landscape connectivity at regional and national scales, and for means of standardizing roadkill data collection and analyses (Evink 2002). Two reports (Evink 2002, TRB 2002a) highlight the need for more systems-level studies addressing long-term issues regarding research in surface transportation and natural systems, in addition to continuing studies focusing on short-term, project-specific, transportation needs.

General sources of research for transportation and ecology include FHWA, TRB, NCHRP, American Association of State Highway and Transportation Officials Center for Environmental Excellence, universities, and other agencies. As discussed above, these efforts have produced a substantial body of literature that documents the effects of roads and traffic on ecological conditions. However, almost no studies of ecological effects of roads have been conducted over long time periods (multiple decades) or at large spatial scales (spatial windows above tens of kilometers). Such studies should be a priority for research. Few, if any, studies directly address ecological effects of road density.

The appropriate scale for research is not always known beforehand, and the ecological impacts of roads can go undetected if an arbitrary scale is chosen for the research. Some multiscale studies have shown that roads affect ecological condition at much larger scales than previously thought. Therefore, multiscale studies can uncover the ecological effects of roads and the scale at which roads affect ecological condition.

Finally, much of the research on the ecological effects of roads can be found in reports that may not have been peer-reviewed or commercially published. For example, committee members are aware of studies documenting the effects of roads on sediment production in reports from the state departments of transportation, the Army Corps of Engineers, and the World Bank. Although included in some searchable databases, such as the Transportation Research Information Service, these reports are not included in scientific abstracting services (for example, Cambridge Abstracts) and, therefore, are generally less accessible to the academic research community. Future studies on the ecological impacts of roads should be published in the peer-reviewed venues.

## SUMMARY

Roads influence ecological conditions across a range of organizational levels and scales. A large part of the scientific knowledge of ecological effects of roads has been based on short-term studies focused on narrowly defined objectives and has generally been related to specific construction or planning needs. As a result, more research is needed on ecological effects that occur over large areas or long periods. Ecological conditions are not only affected by the construction of roads and road appurtenances (bridges and culverts) but also by the traffic on the roads and, at larger scales, by the increases in road density. The ecological effects of roads are much larger than the roads themselves, and the effects can extend far beyond ordinary planning domains. Few studies address the complex nature of the ecological effects of roads. For example, little is known about how roads impede access to foraging areas or key prey species, potentially resulting in cascading or other trophic effects. Studies assessing ecological effects are often based on small sampling periods and, therefore, do not adequately sample the range of variability in ecological systems. More research should be directed at identifying the appropriate scale at which roads affect ecological conditions.

Information on the resiliency of biodiversity components to road-related disturbances is needed to better understand the effects of roads on ecological systems. Research on the ecological effects of roads over long periods or at large spatial scales and research on the complex nature and impacts of roads within ecologically defined areas, such as watersheds, eco-regions, or species' ranges, should be a priority. Research on the local scale should continue, however, because the context of many transportation decisions is at the local scale with direct application, and studies that address the context are likely to be the most frequently used and have the largest influence.

# 4

# Ameliorating the Effects of Roads

## INTRODUCTION

Although the development and operation of the transportation system can affect ecosystems and their components at many scales and in many ways (see Chapter 3), many approaches have been developed to avoid and reduce these effects. Protecting natural resources and providing safe and effective transportation are in the public interest, and addressing both of them is good public policy and practice. Ameliorating environmental effects can occur at all phases of road projects—from planning and design through construction and maintenance operations.

The most protective approach is to avoid environmental effects altogether. Another approach is to take measures to minimize effects that cannot be avoided, and a third approach is to compensate for effects. However, it is useful to consider mitigation as a three-tiered hierarchy involving avoidance, minimization, and compensation in a priority order that reflects the principle that it is generally better to leave something alone than to try to fix it after environmental effects have occurred. Discussions of project mitigation often focus on compensatory mitigation, such as mitigation of on-site impacts or compensation through off-site third-party alternatives (mitigation banks or in-lieu fee programs). The focus on compensatory mitigation options, however, can overlook other important ways that environmental factors can be addressed in transportation work. Understanding mitigation as a sequence is an important part of addressing effects comprehensively. This concept can be followed throughout the process of transportation project planning and development.

Managers have a number of opportunities to improve environmental conditions after design and construction phases of road projects. As roads are built and used, new and unforeseen effects often appear. Some of these effects can be addressed during ongoing maintenance operations on the roadbed and roadside.

This chapter discusses opportunities for ameliorating the effects of roads. It begins with a discussion of mitigating effects organized by scales of administrative organization; the largest scale pertains to national or regional perspectives, the medium scale includes state or highway corridor planning, and the smallest or finest scale applies to the decisions and opportunities associated with individual transportation projects. The chapter ends with a section on opportunities for increased environmental stewardship during routine maintenance operations after construction.

## SCALE-BASED CONSIDERATIONS OF ENVIRONMENTAL MITIGATION

### National or Regional Perspectives (Broad Scale)

At the national and regional scales, opportunities for ameliorating the environmental effects of transportation come in the form of broad public policy for governmental agencies. Environmental issues are often treated as a permitting issue rather than a dimension of project design. Environmental regulations and permits are intended to protect certain types of natural resources, such as wetlands or threatened and endangered species, but this system does not always promote the best, most comprehensive treatment of environmental issues. Actions at the federal level can help to improve the process in the following ways:

• Provide policy, guidance, and funding for transportation design and decision making that take ecological processes into account.
• Expand the knowledge base for assessing potential effects of transportation activities through nationally funded research projects.
• Encourage cross-disciplinary dialogue between engineers, ecologists, and other environmental professionals to raise mutual awareness of each other's expertise, needs, and challenges.

• Share information from practical experience. The larger perspective of national agencies and organizations permits identification and promotion of positive examples of success as well as lessons learned.

## State and Highway Corridor Considerations (Medium Scale)

Opportunities to ameliorate the ecological effects of transportation activities at the medium scale equate best to the planning stages of individual projects, larger road corridors, and statewide transportation system plans. These plans identify long-range needs for meeting transportation objectives.

In medium-scale planning efforts, it is possible to set a direction that can avoid and minimize many ecological effects of roads before projects are planned. In most cases, transportation projects consist of improvements to existing systems rather than consisting of new-road construction options. There are fewer options for minimizing effects in improvement projects; however, improving culverts and bridges and the routing are some ways to reduce the ecological impacts of projects on existing systems. Where new alignments are being considered, changes to the overall pattern of roads in a local area can also be considered. For example, roads may be consolidated or realigned to avoid an ecologically sensitive area, such as a wetland. Other examples are provided in the following section.

Regional planning activities typically occur within politically defined areas, such as municipalities, counties, or portions of states. These boundaries make sense when considering the interaction between socioeconomic systems and a regional plan. However, the interaction between ecological systems and a regional plan is most effectively considered within an ecologically defined area, as suggested in Chapters 3, 5, and 6. The appropriate ecologically defined area will depend on the ecological conditions of interest. For example, if water quality is the environmental issue of concern, the watershed would be the most appropriate planning area. If the concern is for persistence of a particular species or population, the range of that species or population should define the planning area. If the concern is for a particular ecosystem process, the appropriate planning area might be an eco-region. Planning boundaries for assessing the interaction between the road system and the ecological systems should coincide with ecologically defined boundaries, such as watersheds, eco-regions, and species ranges, determined by the complexities of the environmental issue. Multiple concerns should be addressed at the most appropriate planning area and may involve more than one. For example, the appropriate planning area for concerns about the continued survival of a species that is dependent on a particular ecosystem process (improved water quality) may need to consider two ecologically defined boundaries: watershed and species range.

The following discussion on strategies to reduce the effects of roads and traffic considers only the needs of flora and fauna. The committee realizes that transportation authorities must consider and balance multiple interests but offers these strategies to provide guidance on setting priorities for the ecosystem independent of other competing factors.

One way to reduce the ecological effects of roads is to reduce the extent of road corridors in a region, especially through valuable habitat (Figure 4-1). In considering planning options for reducing or limiting the ecological effects of roads, the regional planner is required to first consider whether it is possible to avoid building new roads. In some cases, removal of a road may also be considered. It is usually impractical to remove a road that local residents and businesses depend on, even when substantial ecological benefits may be gained. In addition, road removal is costly. Therefore, road removal is usually not considered to be a practical option.

Many ecological effects of roads are due to traffic rather than to the road itself. For example, pollution from vehicles may alter plant communities to distances of at least 200 m from the road (Trombulak and

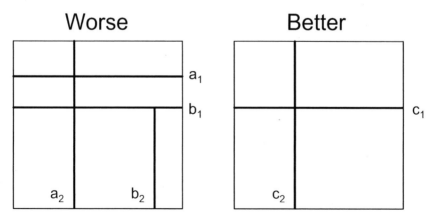

**FIGURE 4-1** Illustration of a reduction in total road length reducing the ecological effect of roads and traffic ($a_1 + b_1 < c_1$; $a_2 + b_2 < c_2$).

Frissell 2000). Increased traffic can also cause an increase in habitat loss because the distance maintained by animals to avoid a road (the road-avoidance zone) increases with increasing traffic volume (Figure 4-2).

There are many possible methods for reducing the total amount of traffic on the roads and concomitant environmental effects. The details of how to achieve these results are outside the charge of this committee and not addressed in this report.

Natural bodies of water, such as rivers, streams, ponds, and wet-lands, are particularly sensitive to the effects of nearby roads and traffic. As discussed in previous chapters, roads affect runoff, especially storm-water runoff, and hence have effects on the flows of nearby streams, on rates of groundwater recharge, and on water quality. For new and recon-struction projects that receive federal funding, the Federal Highway Ad-ministration (FHWA) requires that "standard management practices" be used to control stormwater. For example, stormwater impacts impacts may be managed by routing the water to infiltration areas, away from the road base, to reduce the amount of pollutants discharged to water bod-ies. Stormwater management is still a major concern for projects not on the primary road system (for example, local roads under different jurisdictions).

Natural water bodies are a critical component of the life-history re-quirements of many species, which may be required to move between the water body and other habitat areas on a daily basis or during an annual migration. In mountainous areas of the western United States, roads are intimately associated with major streams and rivers, usually paralleling them in, or next to, riparian areas, which are even more important for

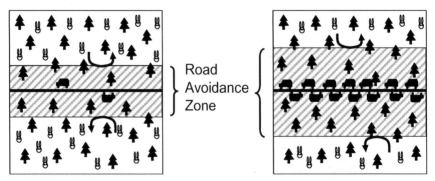

**FIGURE 4-2** Illustration of the effect of traffic volume on the road-avoidance zone or the distance maintained by those animals that avoid a road.

wildlife habitat and movement in arid areas. Roads near water bodies cause high mortality of animals moving to and from the water. For animals that avoid roads or traffic, the presence of a road near a water body can limit the animal's ability to access the water or the other necessary habitats. Roads also increase human access to natural water bodies, resulting in their indirect degradation. Research by Carr and Fahrig (2001) showed that negative effects of road density on aquatic animal populations occur within 1.5 km of ponds and wetlands for turtles and within 2 km of ponds and wetlands for amphibians.

In most cases, it will not be possible to either remove roads or reduce overall traffic volume in the region. However, planners can strive to reduce the ecological impacts of roads by implementing measures that redirect most traffic away from ecologically sensitive natural areas in the region (Figure 4-3).

The previous suggestions illustrated in Figures 4-1 to 4-3 are aimed at reducing or at least not increasing the road density and traffic volume regionally and locally. However, the regional planner is often faced with a demand for increased traffic capacity, which will adversely affect ecological conditions. There are some general strategies that planners can use to minimizing these effects.

In general, there are fewer ecological effects from one road than from two roads. As discussed above, roads themselves (independent of the traffic on them) have ecological effects, such as hydrological changes. The road-avoidance zone (Figure 4-2) is one of the factors con-

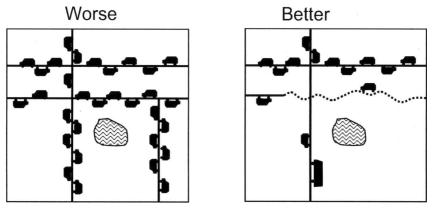

**FIGURE 4-3** Illustration of reducing traffic near ecologically sensitive water bodies.

tributing to the road-effect zone, which widens with increasing traffic (Forman et al. 2003, NRC 2003). However, the width of the road-effect zone increases at a declining rate with increasing traffic (Figure 4-4; Foppen and Reijnen 1994, Reijnen and Foppen 1994, Reijnen et al. 1995, Forman and Deblinger 2000). Measures that reduce roads and traffic within 2 km of water bodies are particularly effective in ameliorating ecological effects of roads. When an increase in traffic must be accommodated, the ecological effects will be less when adding to an existing road rather than building a new road, all else being equal.

All the suggestions illustrated in the figures depend on the particular circumstances under consideration. For example, if the existing road is near a water body, more ecological damage might be caused by increasing traffic on it than by building a new road if it is in a less ecologically sensitive location. There also can be social and political reasons for not increasing traffic volume on existing roads. When more traffic must be accommodated but the volume on existing roads cannot be increased, construction of a new road may be the only viable option. The ecologi-

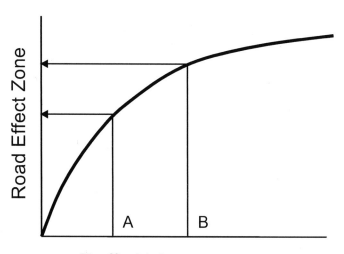

**FIGURE 4-4** Conceptual illustration of the general relationship between the width of the road-effect zone and the volume of traffic on the road. The width of the road-effect zone increases at a declining rate with increasing traffic. For example, traffic volume doubles from A to B, but the road-effect zone increases by much less.

cal effects of a new road can best be minimized by building it as far from natural areas as possible and through developed areas, such as urban areas or areas of intensive agriculture (Figure 4-5). In fact, if a road is placed through a high-intensity agricultural area, roadside verges of seminatural habitat can represent an ecological benefit of the road.

In summary, fewer larger, high-volume roads are preferred to many smaller, low-volume roads, because the impact of roads on the local ecology increases with distance and volume at a declining rate. Another benefit to this strategy is the preservation of larger habitat areas and less habitat fragmentation. Research on rural roads in The Netherlands indicates that habitat fragmentation can be reduced by using techniques for easing the impact of traffic in residential areas, referred to as "traffic calming" (Jaarsma 1997, Jaarsma and Willems 2002). Specifically, traffic calming is used to concentrate traffic flows so that the majority of the traffic is on major roads. The result of these tactics is less construction of new roads and therefore reduced habitat fragmentation, suggesting that a less dense road network with higher volumes results in less habitat fragmentation.

Sometimes, political and economic conditions might not allow implementation of measures to reduce traffic on roads near natural areas. A new road might even be built near or through such areas. Measures can

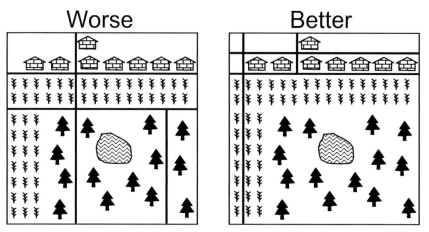

**FIGURE 4-5** Illustration of how the ecological effects of a new road can be mitigated by consideration of surrounding land use. Roads inserted in urban or agricultural areas may have a smaller ecological effect than roads inserted in natural areas.

be implemented that permit water and animals to pass safely under the road, such as lengthening bridges or constructing viaducts, or that permit animals to move over the road, such as wildlife overpasses (Figures 3-4, 4-6). For example, where roads cross small streams, culverts can be replaced by bridges that are long enough and high enough to permit movement of stream animals, as well as terrestrial animals, under the road. Wildlife overpasses also can be used to connect otherwise disconnected populations. Fencing along both sides of the roads discourages animals from attempting to cross the roads and funnels them to the underpasses or overpasses where they can safely cross.

### Using Environmental Data in Medium-Scale Planning Efforts

Factoring ecological information into transportation planning relies on quality information about the environmental resources in the planning area. Due to the broad landscape context of road systems, landscape patterns and processes must be incorporated into the planning and construction process (Forman 1987). When used in a geographic information system (GIS) environment, regional or landscape habitat connectivity models can facilitate decision making in identifying, setting priorities

## Worse                                    Better

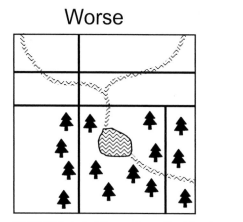
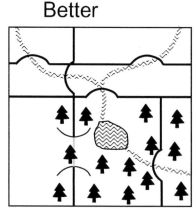

**FIGURE 4-6** Illustration of mitigation measures to reduce the ecological effects of roads when it is not possible to redirect traffic away from natural areas. Measures include lengthened bridges to permit flows of water and animals under the road and wildlife overpasses to permit animal movements over the road. Crossing rivers at right angles shortens the bridge and the ecological effects.

for, and designing amelioration measures (for example, corridors for animal movement) (van Bohemen et al. 1994, Bekker et al. 1995). Use of such models provides for the development of a more integrated land-use strategy by taking into account different land-management practices and priorities of habitat conservation concerns.

At the medium-scale planning stage, evaluating potential environmental effects can take the form of relatively simple screening of project areas (sensitive wetlands, streams, or habitat for species with special management considerations) and pollution sources (contaminated sites or air quality). Evaluation can also consider adjacent land uses, especially where they involve natural-resource management. Other resources could include information on wildlife and vehicle collisions in the area. Rapid assessment methods using GIS hold promise for streamlining the planning process when the effects of transportation on the environment are considered from the onset. These methods are discussed in detail in Chapter 6.

Governmental agencies and nongovernmental organizations that manage natural resources are involved in many types of conservation planning. Some examples include state biodiversity plans, endangered-species recovery plans, and watershed-restoration plans. Many of these plans are not conducted in coordination with transportation planning information and as a consequence do not take the existing transportation network or potential expansion into account. Land-use and environmental-restoration plans present opportunities to link transportation planning with environmental protection. Linking the plans can help to (1) identify natural resources that should receive special consideration of avoidance and minimization strategies; (2) identify ways for transportation projects to contribute to environmental enhancement as part of future project planning and implementation; and (3) provide direction for the location and design of compensatory mitigation sites so that these sites can help to maximize benefits by supporting larger conservation objectives.

A number of examples of this collaborative planning exist in the United States. These joint efforts provide a good opportunity for addressing a broader set of environmental concerns early enough in the transportation project to shape and modify the project. The California Department of Transportation and the Nature Conservancy have undertaken such collaborative projects (Box 4-1). Other examples of such planning efforts can be found in specific transportation plans at the

---

**BOX 4-1** Caltrans and TNC

The California Department of Transportation (Caltrans) and the Nature Conservancy (TNC) have collaborated on a partnership to minimize the environmental impacts from road projects. Developed in 2002, Caltrans and TNC combined California Transportation Investment System data highlighting existing paved roads, current projects, and planned development for the next 20 years and TNC portfolio sites displaying conservation and biodiversity data at multiple scales. The GIS overlays provide both parties a visual representation of key land areas of concern and possible impacts on individual species that may be occurring now or in the future because of road projects. Caltrans and TNC are discussing how the GIS tool can be implemented in transportation planning within the state.

---

transportation-department websites of Florida (see also Chapter 6) and Washington states.

## Environmental Mitigation at the Individual Project Scale (Small or Fine Scale)

The transportation activities in individual projects on road segments correspond to the finest scale of analysis. At this scale, transportation actions include designing and evaluating alternative designs, obtaining environmental permits, developing project-specific mitigation measures, acquiring right-of-way lands, and constructing the road. Ongoing maintenance and road operations also occur at this scale.

### Mitigation Considerations During Project Design

Environmental issues, needs, or desired conditions need to be known at a fine scale of resolution for factoring into project development. In the early stages of the project design, context-sensitive design and alternative development can be selected to avoid and minimize adverse environmental effects. A simple example would be to increase the slope along the sides of a roadway to minimize wetland fill. Other examples include selection of roadway alignment (such as following topo-

graphic contours rather than cutting across them), use of alternative construction methods, or timing of construction.

Many new and alternative construction techniques have been introduced to minimize environmental effects—for example, noise from pile-
driving, which can be disruptive and harmful to wildlife, can be reduced
by the use of vibratory hammers instead of impact hammers. Underwater bubble curtains can be used to reduce sound-pressure transmission
underwater, which is known to harm aquatic animals (Turnpenny et al.
2003). Fish-startle systems that produce high-frequency signals have
been used to reduce the impacts of aquatic blasting by temporarily relo-
cating fish from a project site (FHWA 2003a). The timing of construction projects can be adjusted to avoid periods that are especially sensitive
for fish and wildlife.

Traffic, however, is the primary source of noise that affects animal
behavior (Forman et al. 2003). Strategies to reduce traffic noise, typically intended for human benefit, include improvements in road surfaces
and vehicles and reducing the volume of trucks in ecologically sensitive
areas. Results of studies in The Netherlands showing that effects of traffic noise can result in decreased breeding-bird densities within 1,000 m
of high-volume roads (Van der Zande et al. 1980, Reijnen et al. 1996)
and other traffic noise-related studies have led the Dutch government to
begin repaving high-volume roads with ZOAB, a noise-absorbing porous
asphalt (Piepers 2001). Similar amelioration measures to reduce traffic
noise are under way in the United Kindgdom where the Highways
Agency's goal is to reduce noise on 60% of their trunk roads (major
highways) during the next 10 years (Price 2003). In Germany, a recently
developed tire can produce half the pavement noise of a conventional tire
at the same price (Carstens 2003).

Other noise-reducing strategies include the use of soil berms (low,
smooth earthen ridges) and vegetation that, unlike noise walls, are not
barriers to wildlife movement. Noise reduction can be further improved
by the combination of narrow soil berms along below-grade roads (Forman et al. 2003).

Wildlife crossings are perhaps the most conspicuous amelioration
measures on roadways. Wildlife passages are found in 23 states, 17 of
which are beginning to systematically incorporate wildlife crossings into
roadway designs (Evink 2002) and across Europe (Damarad 2003).
Some of the most successful examples come from Florida, where 32 underpasses were built along Interstate 75, and from Banff National Park,
Alberta, where 24 crossings, including two overpasses, were installed on

the Trans-Canada Highway. In both locations, the measures were effective in dramatically reducing wildlife and vehicle collisions and barrier effects to animal movement (Clevenger et al. 2002b, Evink 2002).

Central median (or Jersey) barriers are frequently used to separate lanes of traffic and often extend over many miles. These 1-1.5-m high concrete structures are designed to prevent head-on collisions, but they also form a wall that can disrupt wildlife movements (Servheen et al. 1998, Forman et al. 2003). Despite the potential impact on wildlife movements, mortality, and driver safety, the "Roadside Design Guide" does not address this issue (AASHTO 2002b).

As a means of increasing road permeability for wildlife, many state departments of transportation are installing Jersey barriers with a modified design that allows passage of small fauna. However, there is no information regarding how effective these modified barriers are, and there are no standard guidelines regarding their placement. Installation of large wildlife crossings is generally an issue of maintaining habitat connectivity for key and wide-ranging, fragmentation-sensitive wildlife species. For many small and medium-sized mammals, drainage culverts have been found to provide critical habitat linkages (Yanes et al. 1995, Clevenger et al. 2001, Foresman 2003).

Spacing intervals for wildlife crossings vary among environments, species for which crossings are designed, and conservation objectives (which might range from simple genetic interchange to more complex restoration and maintenance of ecosystem processes). Therefore, guidelines for spacing have not been developed.

Some environmental concerns can be addressed through setting design standards for projects. For example, wider bridge spans and reducing or eliminating in-water piers help to limit effects on aquatic systems. Design standards for stream crossings that routinely incorporate in-water piers can help to improve environmental conditions. The periodic reconstruction of highway bridges that span riparian areas is an excellent opportunity to improve wildlife passage along riparian corridors by widening bridge spans or by habitat enhancement.

A series of best-management practices (BMPs) for road construction has been developed for reducing or ameliorating the adverse effects of roads on aquatic ecosystems described in Chapter 3. The BMPs often focus on mitigating effects on fish passage or stream habitat (e.g., Tran-Safety 1997, Moore et al. 1999, Robison et al. 1999, Bates et al. 2003). The techniques, summarized by the National Research Council (NRC 2004), include the following approaches:

- Careful planning of the road's route to keep the road on terrain that is resistant to landslides and erosion and to minimize the number of stream crossings.

- Designing bridges and culverts with hydraulic characteristics that allow aquatic organisms to pass through them in both directions, as appropriate to different life stages.

- Engineering techniques, including the appropriate use of vegetation, to stabilize embankments.

- Managing stormwater runoff to reduce hydraulic connections between roads and streams. To the degree that a road is impervious, as paved roads generally are, stormwater runoff is enhanced and concentrated unless provision is made for adequate drainage. Even unpaved roads enhance and concentrate runoff.

- Controlling soil erosion on newly constructed roads, stream crossings, and related structures.

- Maintaining the crossings regularly to prevent debris and beaver dams from allowing culverts to become clogged. A clogged culvert can turn a road crossing (embankment) into an earthen dam, whose inevitable failure will lead to a torrent of water, debris, and sediment.

Unlike other areas of transportation research, there are virtually no design guidelines for building wildlife-crossing systems (Evink 2002). This area of research is an emerging science. Consequently, there are few published studies to refer to for design criteria for habitat connectivity structures for wildlife (for example, wildlife passages) or guidelines for multiple species or fragmentation-sensitive species (for example, wide-ranging large carnivores). The National Cooperative Highway Research Program is currently funding a project that will be the first attempt to develop a state-of-the-art guidance and a decision-support tool to enable well-founded decision making on habitat connectivity structures for wildlife (TRB 2004).

Some amelioration measures are expensive, making it difficult to decide what actions are appropriate and what level of investment is justified. There are definitive standards for some mitigation, such as placement of noise walls. For many impacts, however, there is discretion and flexibility in deciding how to mitigate. Competing interests, costs, and operational issues must be balanced with environmental needs, so it is important to know what scale of enhancement will provide a meaningful ecological benefit. These decisions become more difficult when basic information is lacking, such as preconstruction criteria needed to evaluate the performance of wildlife-crossing structures.

Mitigation projects should have a priori criteria or performance standards that are agreed upon by all responsible for supervising the implementation and functioning of the mitigation measures (NRC 2001). The standards can be designed with some flexibility (or ranking) in terms of goal attainment and target dates and later be refined and updated if required. A small but growing body of information on methods to measure mitigation effectiveness is becoming available as more mitigation projects are undertaken.

Determining how well amelioration measures perform requires long-term study (NRC 2001). Research needs to be an integral part of a highway mitigation project, even long after the measures have been in place. Mitigation is costly, requiring an important investment of public funds. Transportation agencies and biologists responsible for designing approaches to ameliorate highway effects are hampered by the lack of information on the performance of the measures because few post-construction performance studies have been carried out (Romin and Bissonette 1996). Such studies would provide useful information for future decisions.

Long-term research in Banff National Park, Alberta, has been successful in providing valuable information about performance of a variety of measures (including fencing, warning signs, and reflectors) designed to reduce the effects of the Trans-Canada Highway on wildlife populations (Clevenger et al. 2002b). More than 6 years of research and monitoring of 24 wildlife passages, landscape changes around them, and wildlife populations that the passages are designed to sustain have shown the importance of long-term study. A Transportation Research Board report cited the highway mitigation research in Banff National Park as having worldwide importance and as being one of the most successful research projects for wildlife connectivity (Evink 2002).

Data from Banff suggest that the annual patterns and trends of wildlife use of the passages provide strong evidence that there is a learning curve or adaptation period for all wildlife species regardless of passage type (overpass or underpass). Small sampling time periods, typical of 1- or 2-year monitoring programs, are too brief, can provide spurious results, and do not adequately sample the range of variability in wildlife-crossing-use patterns in landscapes with complex wildlife and human land-use interactions. For wildlife passages to be effective over the long term, they will have to be able to accommodate the fluctuations in species, their demographics, and variances in animal behavior while maintaining viable populations around them. Continuous long-term monitoring of wildlife-crossing structures and wildlife populations is critical for assessing the conservation value of mitigation passages for wildlife.

Wildlife crossings are expensive measures, but the gap in devising cost effective designs and decision-support tools based on ecological and engineering criteria leaves no alternative. Roads do not affect wildlife populations equally (Forman et al. 2003). Road mortality has a more immediate impact on population viability compared with barrier effects on genetic isolation. Thus, different levels of effort are required to assess the ecological benefits of wildlife crossings in regard to restored movement patterns, genetic interchange, or road-kill reduction.

### Compensatory Mitigation

Wetlands are ecosystems that must be specifically addressed under the Clean Water Act (see Chapter 5). A review by the NRC (2001) reported on how effective the "no net loss" rule has been for compensatory mitigation of wetlands permitted under Section 404 of the Clean Water Act and how it may be useful for mitigation of transportation project effects on wetlands. The report stated that wetlands processes should be evaluated in the context of the watershed or region within which a wetland exists to maintain wetland diversity, connectivity, and appropriate proportions of upland and wetland systems needed to enhance the long-term stability of the wetland and riparian systems. The report suggested avoiding such wetland types as bogs or fens that are difficult to restore and paying special attention to riparian wetlands. The report recommended that self-sustaining wetlands be the goal of mitigation or restoration. The report's operational guidelines for self-sustaining wetlands are listed in Box 4-2.

### OPPORTUNITIES FOR ENVIRONMENTAL STEWARDSHIP

Although mitigation of environmental concerns is being addressed at a variety of scales during the planning and construction phases of road development, a tremendous opportunity exists for transportation agencies to become better environmental stewards through ongoing maintenance and operations activities. These opportunities focus on managing vegetation in road corridors and rights-of-way, on watershed management, and on improving other targeted ecosystem processes.

---

**BOX 4-2** Operational Guidelines for Creating
or Restoring Self-Sustaining Wetlands

1.  Consider the hydrogeomorphic and ecological landscape and climate.
2.  Adopt a dynamic landscape perspective.
3.  Restore or develop naturally variable hydrological conditions.
4.  Whenever possible, choose wetland restoration over creation.
5.  Avoid over-engineered structures in the wetland's design.
6.  Pay particular attention to appropriate planting elevation, depth, soil type, and seasonal timing.
7.  Provide appropriately heterogeneous topography.
8.  Pay attention to subsurface conditions, including soil and sediment geochemistry and physics, groundwater quantity and quality, and infaunal communities.
9.  Consider complications associated with wetland creation or restoration in seriously degraded or disturbed sites.
10. Conduct early monitoring as part of adaptive management.

---

## Roadside Maintenance and Management

Twelve million acres of land are in public rights-of-way, an area larger than many states (White and Ernst 2003). Both directly and indirectly, these roadside areas are habitats for nonnative invasive species and provide corridors for the expansion of nonnative species. The social values associated with roadside vegetation have changed over time. Initial focus for roadsides was on "beautification" projects or bank stabilization and led to plantings of exotic grasses and plants, such as kudzu (see Chapter 3). Beginning in the 1970s, these aesthetic goals were supplanted by more ecological goals, although the two are not mutually exclusive. Roadside vegetation management seems to be moving in the direction of native species' restoration and maintenance (Figure 4-7). The state of Iowa's roadside management program is an example where transportation agencies are partnering with other groups to achieve goals of restoring large areas of prairie ecosystem (White and Ernst 2003). The committee heard reports from the New York Department of Transportation of similar opportunities for roadside ecosystem restoration carried out as part of normal maintenance operations. The Washington Department of Transportation has a soil bioengineering program that uses native plants and seeds, which are well adapted to local climate and soil

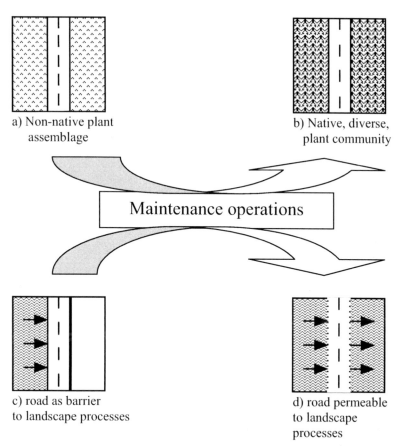

a) Non-native plant
   assemblage

b) Native, diverse,
   plant community

Maintenance operations

c) road as barrier
to landscape processes

d) road permeable
to landscape
processes

**FIGURE 4-7** Examples of ecological restoration performed by ongoing road maintenance operations. The top of the diagram (a to b) indicates restoration activities done in a variety of states, where alien pest plants (a) were replaced by diverse native plants (b). In the lower half of the diagram, many restoration projects have reestablished landscape processes, such as water flow or animal migration.

conditions, to provide erosion control, slope and stream bank stabilization, wildlife habitat, and other benefits. These projects usually require less heavy machinery and can be installed when the site problem is small and during slow construction periods, thereby costing less and causing fewer impacts. The development of training manuals and the sharing of experiences, data, and information would help to build a body of knowledge on ways to achieve these ecological goals.

## Improving Specific Ecosystem Processes

The committee also uncovered numerous examples of small-scale projects that produced environmental benefits. These ranged from sensitive-species mapping and planting, stream habitat improvement, construction of stormwater treatment facilities, and improved aquatic habitat. Other small-scale projects focused on endangered species recovery and restoration, such as riparian and stream restoration for salmon in the Pacific Northwest and fencing for protection of the desert gopher tortoise. Other projects included attempts at decreasing habitat fragmentation by construction of multifunctional crossings, widening of bridges, and improving wildlife connectivity. These projects integrate a variety of objectives, such as improving fish passages and stormwater and noise management, in environmental retrofit programs and can lead to broader-scale, such as watershed scale, improvement of ecosystem functioning (Figure 4-7c,d).

## SUMMARY

This chapter has presented a set of existing and potential opportunities for mitigating the ecological effects associated with three major phases of road projects: planning, construction, and maintenance. At the national scale, the types of conclusions made in Chapters 3, 5, and 6 are all germane. Environmental issues should be addressed earlier in the planning and design phases. Interdisciplinary teams could be developed to share data, understanding, and expertise. At the medium and fine scales, many designs, activities, and projects have been done to mitigate effects. However, information, studies, and databases on the long-term functionality of these mitigation efforts are sparse. Long-term studies and a greater set of analytical tools are needed to help inform decision makers about the implementation of mitigation efforts, such as wildlife habitat connectivity structures. These tools could help to assess appropriate locations and design measures for these efforts. Many other opportunities exist for improving environmental conditions in road project operation and maintenance. Projects have addressed improving water quality, aquatic habitat, habitat for species of concern, control of nonnative plants and animals, and reestablishment of habitat connectivity.

# 5

# Legal Context for Planning and Policy

## INTRODUCTION

This chapter addresses the existing legal planning and policy setting for consideration of ecological effects associated with roads in urban and rural locations. The primary focus is on the federal environmental laws that apply to federally funded or approved highways under programs administered by the U.S. Federal Highway Administration (FHWA).

The chapter first summarizes federal transportation laws, describing the transportation planning and project development structure. While these laws establish requirements for federal funding of transportation projects, they also impose a structure on regional and local transportation planning and project development. Transportation planning, largely a local function, is subject to laws and policies that are different from those that apply to development of specific transportation projects.

These are major federal environmental laws that generally apply to transportation projects, but there is no single law of ecological concerns for transportation planning or projects. Rather, ecological issues are covered, in part, under a suite of different federal laws. FHWA maintains an extensive and easily accessible *Environmental Guidebook* (FWHA 2004a) that includes a summary of all environmental requirements, as well as specific regulations and guidance regarding specific environmental resources.

## TRANSPORTATION PROCESSES: HOW TRANSPORTATION IS PLANNED AND PROJECTS ARE DEVELOPED

Transportation planning and project development reflect the desires of communities, including consideration of the impacts of roads on the natural and human environments. Federal transportation law recognizes the important role of the states and local communities in planning, building, and maintaining roads. To understand the legal requirements concerning ecology and the environment, it is necessary to understand the structure for transportation planning and project development. The processes through which a road gets planned, built, and maintained provide opportunities and obligations to incorporate ecological concerns.

There are substantial differences between *transportation planning* and *project execution or development* with respect to the governmental entities involved and their legal obligations for environmental or ecological resources. Transportation planning encompasses the processes under which states, regions, and localities plan their desired regional transportation systems for periods up to and sometimes exceeding 20 years. Although environmental issues must be considered in transportation planning, the only important mandatory requirement concerns conformity with air pollution standards. As described more fully below, the federal government reviews transportation plans solely to ensure protection of air quality, not to protect any other environmental resources, such as water quality.

In contrast, project execution or development includes all the work done to implement a particular road project, such as a new road, road improvement, or other discrete endeavors. Projects are developed and implemented with detailed, site-specific evaluations of engineering, costs, safety, transportation mobility, and environmental concerns. Generally, the state, city, or county government manages a specific project. Most of the federal environmental requirements apply to project development and implementation rather than transportation planning. In project development, various ecological issues are addressed. An excellent summary of the transportation planning process and environmental requirements is found in Blaesser et al. (2003). Additional information can be found in FWHA (2004b).

## FEDERALLY STRUCTURED
## REGIONAL AND LOCAL PLANNING

### Metropolitan Planning Organization

Transportation planning is the province of the metropolitan planning organization (MPO), a nonfederal entity established under federal laws. Despite its title, and despite the focus of this report on non-urban roads, it is appropriate to consider MPOs here, for two reasons. The first is that MPOs do not operate only in urban areas. Indeed, some areas that are within their purview have considerable wildland or ecosystem values worth protecting. As an example, the proposed Inter-County Connector (ICC) in Maryland is within the purview of an MPO, and yet much of the controversy about that proposed highway concerns the effect it might have on wetlands and other undeveloped ecosystems. There are other examples as well. Thus, this discussion of MPOs is in fact relevant to many roads in nonurban—even rural—areas.

The second reason is that for rural locations lacking population density needed to support an MPO, a rural planning organization or the state conducts transportation planning. The rural or state planning in lieu of an MPO involves the same considerations and similar processes as MPO planning.

The MPO is a transportation policy-making organization and is made up of representatives from local government and transportation authorities. MPOs were created under the Federal Surface Transportation Assistance Act of 1973, which required the formation of an MPO for any urbanized area that had a population greater than 50,000. Subsequent transportation legislation further addressed the duties and boundaries of MPOs. For example, MPO duties were modified in the Intermodal Surface Transportation Efficiency Act of 1991 and the Transportation Equity Act for the Twenty-First Century of 1998 (TEA-21), which had enormous impacts on the nation's transportation system. Together, these two laws dramatically increased investment in roads and bridges, spurred a revival of public transportation, and helped to create a more efficient and interconnected roadway system (AASHTO 2002a, 2003; FWHA 2004b). Under current law, the core metropolitan and statewide transportation planning requirements remain intact. The law emphasizes the cooperation of local and state and local officials with transit operators to tailor the planning process in meeting the needs of state and metropolitan transportation (AMPO 2004).

Current federal law requires that the MPO consider seven objectives in transportation planning, among which is protection and enhancement of the environment (see item D in list below). The U.S. Code (23 USC § 134(f) [2003]) states the following:

(f) Scope of Planning Process.—
    (1) In general. —The metropolitan transportation planning process for a metropolitan area under this section shall provide for consideration of projects and strategies that will—
        (A) support the economic vitality of the metropolitan area, especially by enabling global competitiveness, productivity, and efficiency;
        (B) increase the safety and security of the transportation system for motorized and nonmotorized users;
        (C) increase the accessibility and mobility options available to people and for freight;
        (D) protect and enhance the environment, promote energy conservation, and improve quality of life;
        (E) enhance the integration and connectivity of the transportation system, across and between modes, for people and freight;
        (F) promote efficient system management and operation;
        (G) emphasize the preservation of the existing transportation system.

Environmental protection is a specific planning goal; however, it is one of seven broad goals that the MPO must balance. The statute exempts planning authorities from court challenges that allege that the authority failed to follow any of these planning goals. The planning process is also expressly exempt from application of the National Environmental Policy Act (NEPA).

The transportation planning process is linked to the federal funding structure. Federal transportation funding is distributed to states on a population-based formula, and the states decide how to allocate funds to projects within their borders. Thus, transportation planning commences with a state transportation improvement plan (STIP). The STIP provides general guidance for state transportation programs but leaves most choices, such as when, where, and how to pursue particular projects, to the MPO at the local and regional level.

The MPO engages in two major planning exercises on a regular schedule. Every 3-5 years, the MPO must consider and revise both its long-range plan and its short-range implementation plan. More frequent planning is required for localities suffering poor air quality. These planning processes are critical steps during which communities consider their future transportation needs. To qualify for federal funding, projects must arise from properly adopted long-range transportation plans (LRTPs) and STIPs.

## The Long-Range Transportation Plan or Metropolitan Transportation Plan

The MPO (or state for rural areas) adopts the LRTP, identifying the various ways that a region plans to invest in the transportation system over a 20-year period or more. The LRTP includes "both long-range and short-range program strategies/actions that lead to the development of an integrated intermodal transportation system that facilitates the efficient movement of people and goods" (23 CFR § 450.322(a) [1999]). For example, the LRTP may identify specific road segments to be widened, general locations for construction of new roads, major reconstructions or repairs, and other systemic improvements. The LRTP provides a regional, systemwide "blueprint" for the transportation network, sometimes also viewed as a regional "wish list."

The LRTP (23 CFR 450C § 450.322 [1999]) contains several elements, including the following:

- Identify policies, strategies, and projects for the future.
- Determine project demand for transportation services over 20 years.
- Focus at the systems level, including roadways, transit, nonmotorized transportation, and intermodal connections.
- Articulate regional land use, development, housing, and employment goals and plans.
- Estimate costs and identify reasonably available financial sources for operation, maintenance, and capital investments (see Part II, section on financial planning).

- Determine ways to preserve existing roads and facilities and make efficient use of the existing system.
- Be consistent with the statewide transportation plan.
- Be updated every five years or three years in air quality nonattainment and maintenance areas.

MPOs make special efforts to allow interested parties to be involved in the development of the transportation plan. Although governed by local law, MPOs generally have public sessions on LRTPs and must meet mandatory public participation standards. In cases where a metropolitan area does no meet national air quality standards (a nonattainment area) or is now in compliance (a maintenance area), the plan must conform to the state implementation plan (SIP) for air quality.

## Transportation Improvement Program

The MPO (or state for rural areas) also adopts the transportation improvement program (TIP), a 2-year program, under a restricted budget, that covers priority implementation projects and strategies from the LRTP. The TIP identifies what the locality wants to commence in the short-term, 2-year period. The TIP enables a region to allocate its transportation monetary resources between the various operating needs and the capital needs of the area. These allocations are based on a clear set of short-term transportation priorities.

Under federal law, the TIP (23 CFR 450C § 450.322 [1999])

- Covers a minimum two-year period of investment.
- Is updated at least every two years.
- Is realistic in terms of available funding (known as a fiscally constrained TIP) and is not just a "wish list" of projects.
- Conforms to the SIP for air quality if the region is designated a nonattainment or maintenance area.
- Is approved by the MPO and the governor for air quality.
- Is incorporated into the statewide transportation improvement program (STIP).

As with the LRTP, the MPO will provide notice and hold public hearings when adopting a TIP.

## Cooperative Planning

Most MPOs act as an overall coordinator for planning and, by virtue of adoption of the TIP, programming funds for operations and projects. Implementing the projects in the TIP generally becomes the responsibility of the states' Departments of Transportation (DOTs) or the city or county governments. Given the roles of different levels of government, the MPO must involve local transportation providers in the planning process. Thus, during the transportation planning stage, MPOs include a wide range of entities, such as state DOTs, transit agencies, maritime operators, airport authorities, rail-freight operators, port operators, Amtrak, and regional operating organizers.

Transportation planning is a cooperative phase. No single agency has full responsibility for the planning of construction, operation, or maintenance of the entire transportation system. For example, some roads that are part of the interstate highway system are subject to certain federal standards. However, they are usually maintained by a state DOT. Other roads are city streets or county arterials designed, operated, and maintained by local municipalities or counties. Transit systems are often built, operated, and maintained by an entity separate from the highway authority. The MPO is responsible for seeking the participation of all relevant stakeholders and agencies in the planning process.

Funding for roads also depends on actions by state legislatures; even federally supported roads require states to share in the costs. By allocating funds and providing other direction, state legislatures can adjust priorities for particular projects within the planning horizon of the TIP or the LRTP. For example, the legislature may opt to accelerate or postpone funding for particular projects identified in the TIP or LRTP. The LRTP and TIP processes generally result in "wish lists" with shifting priorities.

## Project Development and Execution

After the state planning, MPO long-range planning, MPO short-range implementation planning, and legislative involvement, specific road projects are ready for implementation. At this project development and execution level, most of the laws concerning ecology and environment apply.

Project development involves all the site-specific considerations, including safety, engineering, and environment, that influence decisions regarding how and where to build a project. For example, the TIP may have identified the need to add capacity to a road segment based on projected population growth and assumed that the road should be widened by one lane in each direction. Project development efforts will involve evaluating the various technical issues of such a project. The evaluation will include detailed transportation analysis (needs), structural assessment (engineering), community effects assessment (noise, dislocations, and viewsheds), and safety and environmental (all natural resources) assessments. After these assessments, a proposal (and alternatives) will be considered. Assuming approval, the project will be built and subsequently maintained. Project implementation involves decisions on financing, construction, and subsequent operation and maintenance. During implementation, a number of activities will be subject to specific permit controls and limitations.

As even this short synopsis illustrates, project development and implementation involves the greatest detail of environmental consideration, largely because it is the stage at which the precise physical effects are evaluated and the most specific potential environmental effects are known.

## FHWA Environmental Policies

The surface transportation laws administered by FHWA provide for evaluation of environmental concerns at all levels of transportation planning, development, execution, and operation (23 USC §§ 109(h), 128 [2003]; 23 CFR 771, 772 [2001]). These transportation laws set forth many guiding environmental standards that must be followed by the FHWA and other highway governing bodies when building and maintaining highways. These laws ensure that decisions on highway projects consider the public interest and that all possible economic, social, and environmental adverse effects of highway projects and locations are considered. These laws apply to the planning and the development of any proposed projects on any federal-aid highway system. Consideration of environmental matters allows for balancing of environmental and transportation interests.

FHWA has long maintained strong environmental policies. The FHWA 1994 environmental policy statement (EPS) represents a continued commitment to the enhancement and protection of the environment by providing the policies and procedures for the development of environmentally sound projects (FWHA 2004a). The 1994 policy updated the 1990 FHWA environmental policy. In 2002, the President issued Executive Order 13274, Environmental Stewardship and Transportation Infrastructure Project Reviews, which reiterated the environmental goals of the U.S. Department of Transportation. FHWA environmental policies and guidance are available through the environmental programs of the FHWA Office of Planning, Environment, and Realty.

The EPS provides guidance on integrating environmental concerns into the full transportation process:

> For an effective, environmentally sound transportation system, the Federal-Aid Highway Program and its projects must incorporate environmental considerations and neighborhood and community values and goals into every phase of transportation decision making. But FHWA must practice environmental sensitivity on an even broader scale. Environmental objectives must be considered in every aspect of FHWA's organization and decision-making (FHWA 1995).

The EPS and current FHWA policy also require

- Consideration of social, economic, and environmental issues equally with engineering issues.
- Coordination of planning to conform to air quality implementation plans.
- Defining the "environment" to include the natural and built environment.
- Encouragement of broad-based public involvement early and continuously in the process.
- Encouragement of continual consideration of environmental factors throughout all phases of project.
- Encouragement of corridor preservation to ensure early consideration of environmentally sensitive areas.
- Encouragement of enhancement of the natural and human environment.
- Encouragement of mandates going beyond compliance to strive for environmental excellence.

- Integration of environmental goals and effects by local, regional, and state land-use planning.
- Promotion and support of watershed planning.
- Promotion of multi-modal solutions to transportation and air quality problems.
- Requirement of environmental commitment compliance and implementation.
- Requirement of full compliance with environmental protection laws, regulations, Executive Orders, and policies.
- Requirement of full consideration of avoidance, minimization, and mitigation of adverse effects.
- Support for an interdisciplinary approach.
- Support for environmental training to develop environmental professionals in transportation.
- Support for research and development to raise the level of expertise to state of the art.
- Support for the merger of NEPA with other environmental reviews and decisions regarding permits.

Additional FHWA environmental policy provides that

- To the fullest extent possible, all environmental investigations, reviews, and consultations be coordinated as a single process, and compliance with all applicable environmental requirements be reflected in the environmental document required by this regulation.
- Alternative courses of action be evaluated and decisions be made in the best overall public interest based upon a balanced consideration of the need for safe and efficient transportation; of the social, economic, and environmental effects of the proposed transportation improvement; and of national, State, and local environmental protection goals.
- Public involvement and a systematic interdisciplinary approach be essential parts of the development process for proposed actions.
- Measures necessary to mitigate adverse effects be incorporated into the action. Measures necessary to mitigate adverse effects are eligible for Federal funding when the Administration determines that

—The effects for which the mitigation is proposed actually result from the Administration action; and

—The proposed mitigation represents a reasonable public expenditure after considering the effects of the action and the benefits of the proposed mitigation measures. In making this determination, the Administration will consider, among other factors, the extent to which the proposed measures would assist in complying with a Federal statute, Executive Order, or Administration regulation or policy.

- Costs incurred by the applicant for the preparation of environmental documents requested by the Administration be eligible for Federal assistance.

The national policy Executive Order 13274 declares,

The development and implementation of transportation infrastructure projects in an efficient and environmentally sound manner is essential to the well-being of the American people and a strong American economy. Executive departments and agencies (agencies) shall take appropriate actions, to the extent consistent with applicable law and available resources, to promote environmental stewardship in the Nation's transportation system and expedite environmental reviews of high-priority transportation infrastructure projects.

Among the steps taken to implement this policy, the FHWA adopted the "Vital Few Environmental Goals" to focus some of its efforts on environmental stewardship. This policy has several specific objectives (FHWA 2004c):

Objective 1
*To improve the environmental quality of transportation decision-making, all 50 States, the District of Columbia, Puerto Rico, and the Federal Lands Highway (FLH) Divisions will use, by September 30, 2007:*

- *Integrated approaches to multimodal planning, the environmental process and project development at a systems level; and/or*
- *Context Sensitive Solutions (CSS) at a project level.*

Objective 2
*To improve the timeliness of the both the federal aid and FLH environmental process:*

- *Establish time frames for EAs [environmental assessment] and EISs [environmental impact statement] and meet the schedules for 90% of those project by September 30, 2007;*
- *Decrease the median time it takes to complete an EIS from 54 months to36 months by September 30, 2007; and*
- *Decrease the median time to complete an EA from approximately 18 months to 12 months by September 30, 2007.*

Objective 3
*To increase ecosystem and habitat conservation, implement by September 30, 2007, a minimum of 30 exemplary ecosystem initiatives in at least 20 States or Federal Lands Highway (FLH) Divisions.*

Objective 1 encourages consideration of environmental concerns at the planning as well as at the project setting. Context-sensitive solutions or context-sensitive design is a policy to enhance environmental issues at the project stage and allows the setting to help guide how the projects are designed (Box 5-1).

Objective 2 addresses time frames and reduction of delay. The issue of whether projects are delayed because of environmental compliance obligations and the need for "streamlining" is complex and is related to the quality of available environmental information addressed elsewhere in this report.

In furtherance of objective 3, in October 2003, FHWA (2003b) released criteria for evaluating projects that increase ecosystem and habitat conservation. FHWA has publicized projects that exemplify ecosystem and habitat conservation on its website and in conferences and training. These policies encourage attention to and protection of ecosystem functions at all levels of transportation.

To implement its environmental policies, FHWA has developed numerous guidance documents, training, and other information to assist states and localities. These are available through FHWA and other transportation agencies. For example, with respect to wildlife and roads, FHWA offers a website called Critter Crossings, providing links and guidance for wildlife crossings. The program describes working examples of wildlife protections in road systems. FHWA also offers guidance on the ecosystem approach and transportation and information on exemplary programs (FHWA 2003b,d). FHWA administers its policies regarding ecosystem analysis and protection through guidance and training.

---

**BOX 5-1** Context-Sensitive Solutions and Designs

"Context sensitive design asks questions first about the need and purpose of the transportation project, and then equally addresses safety, mobility, and the preservation of scenic, aesthetic, historic, environmental, and other community values. Context sensitive design involves a collaborative, interdisciplinary approach in which citizens are part of the design team."

Thinking Beyond the Pavement, Maryland State Highway Administration Workshop, 1998 (MSHA 1998)

The following case study illustrates how context-sensitive solutions or designs can foster improved relations between all interested parties, while encompassing environmental, historical, and scenic conservation or mitigation, and still maintain the overall goal of providing safe and efficient transportation. The case study will focus primarily on the environmental aspects even though the project contains other context-sensitive solution factors.

The Smith Creek Parkway project (Martin Luther King Jr. Parkway) in Wilmington, North Carolina, consisted of an area of more than 7 miles and was divided into four sections. Two of the sections consist of wetlands. The project was enacted to reduce traffic congestion, travel times, and congestion-related accidents and to provide a link between U.S. 74 and downtown Wilmington. The coordinated effort during the development phase of the project consisted of several agencies, local government officials and agencies, special-interest groups, local businesses, and the citizens of Wilmington.

The most important environmental issue of the project was the preservation of the surrounding wetlands. Originally, a box culvert was proposed, but after the 1992 NCDOT feasibility study, a large bridge spanning over the wetland area provided a more suitable solution. The roadway section was reduced from 14.4 acres to 5.35 acres, reducing the number of lanes from six to four. Further, compensatory mitigation was used to restore tidal swamp forest adjacent to the wetland. Possible contamination of the wetlands by hazardous waste and hazardous materials in the vicinity was avoided by shifting the project to the north of the originally proposed location. Studies showed no threatened or endangered species were found within the project area.

Although the overall project duration was long and the original design was changed, the coordination of all interested stakeholders under a context-sensitive design, reduced ecological impacts, while providing transportation benefits to the residents of Wilmington.

## MAJOR FEDERAL ENVIRONMENTAL LAWS

A number of federal laws and regulations require consideration of ecological issues or protection of ecological resources before, during, and, to a degree, after road construction. For the most part, federal law requires consideration of a broad range of environmental aspects, including wildlife and ecological effects, when a federal agency either undertakes or funds a road project. Some laws protect specific ecological resources with requirements to preserve or protect the resource by avoidance or compensation for effects. This section addresses selected major federal statutes through which ecological concerns of roads are or could be addressed.

Most of the laws summarized here are not limited to or directed at road projects, nor are they focused on ecological concerns. Rather, ecological issues are captured in some way through implementation of these laws. Unless otherwise specifically addressed, these laws apply only to project development and execution, not to transportation planning.

### National Environmental Policy Act

Known as the "granddaddy of environmental law," the NEPA (42 USC §§ 4321-4335 [2003]) requires federal agencies to describe and disclose a wide range of environmental consequences of actions they undertake or fund, including ecological information. Section 101 (Appendix C) sets forth the lofty goal and policy of NEPA that the federal government "use all practicable means and measures" to, among other goals, "attain the widest range of beneficial uses of the environment without degradation, risk to health or safety, or other undesirable and unintended consequences." The heart of NEPA is Section 102(2)(C), which provides that each agency include in every recommendation or report on proposals for legislation and other major federal actions significantly affecting the quality of the human environment, a detailed statement by the responsible federal official on—

(i)    the environmental impact of the proposed action,

(ii)   any adverse environmental effects which can not be avoided,

(iii)  alternative to the proposed action,

(iv)   relationship between local short-term uses of man's environment and enhancement of long-term productivity, and

(v)    any irreversible or irretrievable commitments of resources.

Pursuant to this section, the Council on Environmental Quality (CEQ) and the FHWA issued regulations describing the process for the required "detailed statement" (40 CFR 1500-1508 [2003] [CEQ]; 23 CFR 771, 772 [2001] [FHWA]). Almost all road projects must be analyzed with either an EIS or an EA that considers a broad range of environmental and ecological factors (23 CFR 771 [2001]).

The CEQ regulations provide overall guidance for applying NEPA, and the FHWA regulations add terms that apply specifically to road projects. Both sets of regulations require that the federal decision maker examine the alternatives to and the effects of the proposed project (40 CFR §§ 1502.1, 1508.9 [2003]). For road projects, alternatives most frequently take the shape of different modes of transportation (such as trains and buses), different configurations (such as fewer lanes), and different locations for the project. Additionally, the no-action alternative—not building the road—must always be examined (40 CFR § 1502.14(d) [2003]).

The environmental document (EIS or EA) must examine the effects of the reasonable alternatives to the project on a variety of resources, including ecological resources. FHWA requires examination of effects to air quality, water quality, noise receptors, wetlands, wildlife, floodplains, wild and scenic rivers, coastal resources, threatened or endangered species, historic and archeological resources, visual resources, land use, farmland, society, the economy, joint development, and pedestrians and bicyclists (FHWA 1987). The EIS or EA is made available to members of the public, and comments are received and considered (40 CFR §§ 1502.19, 1503 [2003]; 23 CFR §§ 771.119(d), 771.121, 771.123(g), 771.125 (a)(g) [2001]).

NEPA is not an action-forcing statute; it only requires that all environmental effects be considered, not that any particular outcome be accepted. In other words, the decision maker need not choose the most environmentally beneficial alternative (although other laws may so require.) Thus, NEPA does not protect ecological resources but merely educates decision makers about ecological resources that could be affected by a proposed action. However, in practice and in conjunction with other action-forcing laws, NEPA often results in the selection of less ecologically harmful alternatives and the mitigation of certain adverse effects. Following preparation of the NEPA document, FHWA announces its decision in a record of decision (ROD) (23 CFR § 771.127 [2001]), which may include environmentally protective conditions.

NEPA policy requires that an EIS identify proposed mitigation for all significant adverse effects identified in the process. An FHWA ROD following an EIS will describe the significant effects identified and the mitigation for those effects. The mitigation conditions become an enforceable part of the federal funding agreement. As a matter of policy, therefore, FHWA uses NEPA to mitigate adverse environmental effects. Frequently, the effects and mitigation terms involve wildlife or other ecological resources.

## Clean Water Act

Aquatic ecosystems, including rivers, lakes, streams, coastal locations, and wetlands, are protected from degradation under programs of the Clean Water Act (CWA). The statute establishes three programs relevant to highway effects. Section 402 addresses direct discharge of pollutants, including highway runoff, into waters. Section 404 addresses dredging and filling of waters and wetlands. Finally, nonpoint source pollution is also addressed under the CWA.

The CWA is administered in a delegated program system under which the U.S. Environmental Protection Agency (EPA) delegates most authority to the states, subject to federal oversight. In practice, therefore, most water protection activity occurs at the state level. In addition, any federal action that might affect a state's water quality must obtain a state water quality certification under Section 401 (33 USC § 1341 [2003]). The state must certify that issuance of the proposed federal permit will not result in any violations of state water quality standards.

### Section 402 National Pollutant Discharge Elimination System Program

Under the CWA national pollutant discharge elimination system (NPDES) Program, discharges of pollutants, including stormwater systems, are regulated as point sources under programs developed pursuant to Section 402 (33 USC § 1342 [2003]). EPA and the states have extensive regulatory programs requiring permits for construction and operations that will result in pollutants entering the nation's waters. NPDES permits are regularly issued to control turbidity (silt) or other waterborne pollutants from road projects into waterways. Discharges during road construction and after a road is built will require NPDES permits, generally issued by the state under EPA authorization.

Covered discharges are those that flow into "navigable waters" (33 USC § 1362(12) [2003]). In turn, "navigable waters" are all "waters of the United States" (33 USC § 1362(7) [2003]). The navigable aspect of the regulated waters has not been viewed strictly, having been expanded to nonnavigable tributaries (33 CFR § 328.3(a)(5) [2002]). There remains active litigation over whether certain isolated or low flowing (ephemeral or intermittent) waters are federally regulated; the general rule is that any discharge into any waters will require a permit.

The water discharge permit will impose standards designed to protect the water quality of the stream, river, or other receiving waters. The standards are not established or revisited for each road project. Rather, the standards are set through separate processes (rule makings) that consider the state preferences for use of a water segment (for example, how clean the state wants the water to be). Standards are also based on scientific understanding of the public health and environmental consequences of pollutant discharges. Aquatic life is considered in setting water pollution standards, which are subsequently included in NPDES permits.

States have primary responsibility to designate the desired water quality for water bodies. States may classify waters (for example, streams or stream segments) based on aquatic habitat, such as fisheries. Once designated, water quality standards are set to meet and protect the designated users. These water quality standards are then made part of NPDES permits. The permit requirement protects aquatic resources, including fish habitat, and consequently the broader functions served by the waterway in the ecosystem.

**Section 404 Dredge and Fill Permits**

Discharge of dredged or fill material is regulated under Section 404 (33 USC § 1344 [2003]). The U.S. Army Corps of Engineers (Corps), rather than EPA, is authorized to issue permits allowing placement of fill in waters, including wetlands. The Corps and EPA have extensive regulations establishing the environmental standards for Section 404 permits (33 CFR 320, 330 [2002]; 40 CFR 230 [1999]). The wildlife habitat function of wetlands is specifically recognized in these regulations.

Wetlands, which vary from often-submerged riparian zones to remote features surrounded by upland, support an enormous variety of wildlife, from fish and insects to large mammals and ungulates. Wetlands are considered "special aquatic sites" under Section 404 regulations. Special aquatic sites are ecosystems of significant biological or resource value that have additional protection under the regulations. For projects that do not need to be located on or in waters, an applicant for a Section 404 permit in wetlands must demonstrate that there are no practicable alternatives to the proposed filling of wetlands (40 CFR § 230.10(a) [1999]). Thus, proponents of road projects that will affect wetlands must meet a significant burden to show that there are no alternatives.

The Section 404 nationwide permit (NWP) program authorizes certain activities that allow small effects in wetlands to occur with less administrative burden than required for an individual permit. For road crossings with small effects (those less than one-tenth of an acre for an entire project), activity can go forward without notifying the Corps under NWP 14 (67 Fed. Reg. 2036 [2002]). Slightly larger road-crossing effects, those of 0.5-acre or less, are eligible for a NWP 14, but the project proponent must notify the Corps before construction and will probably be required to mitigate the effects.

In addition to the nationally applicable NWPs, local Corps districts may have locally applicable general permits under which permit approval can be given to broad categories of activities that have small effects in wetlands. For example, Virginia has a state program general permit (SPGP), which allows for certain limited discharges associated with linear transportation projects. The SPGP establishes application, restoration, and mitigation requirements for various quantities of effects and attaches numerous general and special conditions to the SPGP's use (DEQ 2004).

Almost all wetland permits require mitigation for adverse effects. The mitigation analysis involves a sequence of steps in which the permit applicant must demonstrate (1) avoidance of wetland effects, (2) minimization of wetland effects, and (3) compensation for unavoidable wetland effects. Wetland mitigation policy is set forth in a 1990 interagency memorandum of agreement between EPA and the Corps, in a determination of mitigation under the CWA Section 404(b)(1) guidelines (effective February 7, 1990), and in agency regulations and other guidance to the field. On December 24, 2002, the Corps provided additional information for compensatory mitigation projects (EPA 2004, USACE 2002).

One means of providing mitigation is by purchasing credits from a "wetlands mitigation bank." Congress expressed support for wetlands mitigation banking by stating that transportation projects should look to mitigation banks to provide compensatory mitigation (23 USC §§ 103(i)(13), 133(b)(11) [2003]). Provisions of the FHWA regulations address wetlands and wetlands mitigation. For example, Title 23, Parts 771 and 777 of the *Code of Federal Regulations* (2001) address use of mitigation banks to provide compensatory mitigation for unavoidable effects to wetlands. Mitigation banking guidance was issued in 1995 by the Corps, EPA, U.S Fish and Wildlife Service (USFWS), and the National Oceanic and Atmospheric Administration (NOAA) (60 Fed. Reg. 58605-58614 [1995]). In July 2003, FHWA, the Corps, and EPA issued guidance for the use of mitigation banking in the federal highway and interstate highway programs (FWHA 2003e).

## Nonpoint Source Pollution

Road projects must also be consistent with state nonpoint-source-pollution management programs developed under Section 319 (33 USC § 1329 [2003]). Nonpoint source pollution, such as sheet flow or general runoff (escaping a storm sewer system or other discrete conveyance), is not regulated with a federal permit. Rather, states are required to plan to address these sources of pollutants. The states can take account of nonpoint source pollution when they establish water quality standards and total maximum daily loads (TMDLs) of pollutants. As a general matter, land-use management controls (for example, keeping the land-surface areas free of pollutants that can wash into water bodies) are the tools for managing nonpoint source pollution. Congress left this authority with states and local governments.

## Rivers and Harbors Act

The River and Harbors Act of 1899 (33 USC § 401 [2003]) protects traditionally navigable waters from "obstructions" to navigation. Any construction, dredging, or filling of navigable waters must be authorized by a permit issued by the Corps. Bridges over navigable waters must be permitted. Corps regulations define navigable waters and describe the permit process. Generally, bridge pilings in navigable waters must be constructed to not interfere with the waters' flow (33 CFR §§ 322, 323.3(c), 329 [2002]). The restrictions imposed under the Rivers and Harbors Act help to maintain the free flow of water, which is an essential component of a healthy aquatic ecosystem.

## Endangered Species Act

Certain species of plants and animals are protected under the Endangered Species Act (ESA). This law prohibits the "taking" of endangered or threatened species, defined as the harassment, harm, pursuit, hunting, shooting, wounding, killing, trapping, capture, or collection of such species (16 USC §§ 1532(19), 1538(a) [2003]). Activity that significantly modifies a species' habitat can be found to "take" the species even if injury to a particular animal or plant is not identified.

Species are made subject to the ESA through a listing process. The U.S. Department of the Interior (DOI) determines which species warrant designation as "threatened" or "endangered" based on scientific data. Listings are subject to public notice and comment before being finalized. When listing a species, USFWS can also determine whether the species depends on a limited, specific habitat and, if so, identify the "critical habitat." The ESA does not authorize a category of protection or listing for healthy species or for declining species that do not meet the status of threatened or endangered.

For any road projects authorized by the federal government that might "take" threatened or endangered species, the ESA requires the authorizing agency to consult with USFWS or NOAA-Fisheries before approving the activity. This step allows USFWS to determine whether the authorized activity is likely to jeopardize a listed species (16 USC § 1536(a) [2003]). This process is known as "Section 7 consultation."

The authorizing agency (such as FHWA) begins the consultation process by asking USFWS to determine whether a protected species is present in the project area. If so, consultation formally begins with the authorizing agency preparing information about the species, known as a "biological assessment," to assist USFWS in making its "jeopardy" determination. USFWS sets forth its conclusions in a "biological opinion" (16 USC § 1536(c) [2003]; 50 CFR § 402.14(h) [2003]). If USFWS concludes in its biological opinion that the project is not likely to jeopardize the recovery of the species, it issues a "no-jeopardy" opinion.

The ESA is designed largely to protect species. The law authorizes the taking of individual animals under certain conditions, as long as the take does not jeopardize the species. Under Section 7, USFWS is authorized to provide an "incidental-take" statement concerning circumstances where the federal action might harm or otherwise take members of a protected species in carrying out a project (50 CFR § 402.14(i) [2003]). For example, a road project near a bald eagle nesting site might disrupt the nesting. The USFWS incidental-take statement can provide terms and conditions to allow but perhaps minimize that take. Without a jeopardy finding at the species level, the presence of listed species does not automatically bring the project to a halt; USFWS can still authorize the activity after requiring compensatory mitigation as a condition for authorization (50 CFR § 402.14(i)(5) [2003]).

If USFWS determines that the project is likely to jeopardize a species' recovery, it issues a jeopardy opinion stating that the project as proposed will violate the act. If appropriate, USFWS may also suggest reasonable and prudent alternatives to the action that will not violate the ESA (16 USC §§ 1536(b)(3)(A), 1536(g), 1536(d) [2003]; 50 CFR § 402.14(h)(3) [2003]). The applicant can accept such an alternative or appeal to the Endangered Species Committee. Nonfederal parties may appeal decisions of USFWS in federal court. Consultation is described more fully in Title 50, Part 402 of the *Code of Federal Regulations* and in the federal *Endangered Species Consultation Handbook* (64 Fed. Reg. 31285 [1999]).

By directly protecting threatened and endangered species from harm, the ESA provides a strong tool to protect certain species. Direct protection of these species or their habitat also provides significant protection to other members of the affected ecosystem. For example, ESA conditions that place restrictions on the reduction of bird habitat in turn leaves that ecosystem available for use by numerous other species as well. Without the presence of a listed threatened or endangered species or their designated critical habitat, the ESA offers no authority to protect ecosystems (NRC 1995).

## Clean Air Act

Designed to protect air quality, the federal Clean Air Act (CAA) may require road projects to obtain permits for air emissions, such as fugitive dust or emissions from machinery. These permits are generally issued by states under delegated authorization from EPA. Although airborne pollutants can affect natural habitat, the CAA standards are largely based on concern for human public health. One notable exception is the acid rain control program, whose goals include protection of forests and fauna (42 USC § 7651 [2003]) (acid rain "presents a threat to natural resources, ecosystems, materials, visibility, and public health") (EPA 2003a).

In addition, the CAA requires consideration of the effects of traffic emissions on regional air quality. This is the "conformity" requirement. New roads must "conform to" the state SIPs, which states design to comply with the federal National Ambient Air Quality Standards (NAAQS). NAAQS are set by EPA, establishing by pollutant the concentration limit in the ambient (outdoors) air to protect the public health. The SIP applies the NAAQS within air basins of a state. Among other things, the SIP sets air emission budgets by pollutant. The operation of a road cannot cause a locality to violate its air quality standards by exceeding the budgets allotted for mobile sources (42 USC § 7506(c) [2003]; 40 CFR 93, A and B [2003]; 40 CFR 51, T and W [1999]).

The CAA requires that each state develop a SIP for pollutants within designated air basins. The SIP explains how the state plans to meet the NAAQS for each type of air pollutant, according to the schedules included in the CAA. Then, the region within the air basin undergoes a regional emissions analysis, which assigns each an emissions budget.

The conformity requirements essentially prohibit federal agencies from supporting any activity that does not conform to a SIP. The proponents of a new or expanded road or of other transportation projects must show that the activity will not (1) cause or contribute to any new violation of any standard in any area; (2) increase the frequency or severity of any existing violation of any standard in any area; and (3) delay the timely attainment of any standard or any required interim emission reductions or other milestones (42 USC § 7506(c)(1) [2003]).

Transportation conformity determinations are conducted for road planning and project development and construction. The conformity analysis looks at a future time—target period for the plan or project—and estimates compliance with air quality standards. As a general rule, the following actions require transportation conformity determinations: (1) the adoption, acceptance, approval, or support of  LRTPs, LRTP amendments, TIPs, and TIP amendments[1] developed by a MPO or the DOT; and (2) the approval, funding, or implementation of FHWA or Federal Transit Administration (FTA) projects (42 USC § 7506(c)(2) [2003]). "Regionally significant" transportation projects carried out by recipients of federal funds designated under Title 23 of the *U.S. Code*

---

[1] As described above, LRTPs are long-range transportation planning documents examining a period of 20 years or more (49 USC § 5303(f)(2) [2003]); TIPs are short-term plans specifying the planning goals for the upcoming years (49 USC § 5304(b)(1) [2003]). Individual projects are described in both LRTPs and TIPs.

and the federal transit laws also require a conformity analysis (40 CFR §§ 93.102(a) [2003]).

The MPO makes the initial conformity determinations when it develops or revises a TIP or LRTP. That determination must be approved by FHWA. Although it is not required, many MPOs develop public education and communications programs that inform the public of the connection between transportation and air quality in their respective regions. These programs also encourage the public to make travel choices that will benefit air quality. MPOs can participate in air quality planning to identify transportation programs and projects that help to reduce emissions from on-road mobile sources of pollution.

Below are several ways to help meet emission-reduction targets for on-road mobile sources:

• Vehicle emission programs (for example, the use of reformulated gasoline or implementation of inspection and maintenance [I/M] programs).

• Changing ways to travel (for example, ride sharing or use of transit).

• Transportation projects that reduce congestion (for example, signal synchronization programs).

Generally, if a road project will have emissions consistent with the motor-vehicle emission budgets of the relevant SIP, it satisfies the transportation conformity standards. However, no individual transportation project can be approved, even if it is in conformity with the applicable SIP, if the LRTP and TIP are not conforming at the time of project approval (40 CFR § 93.114 [2003]). In general, no transportation project may cause or contribute to any new violations of localized concentrations of carbon monoxide (CO) or particulate matter up to 10 micrometers in diameter ($PM_{10}$) or increase the frequency or severity of any existing violations of CO or $PM_{10}$ concentrations in CO or $PM_{10}$ nonattainment and maintenance areas (40 CFR § 93.116(a) [2003]).

Motor-vehicle emission budgets can be revised, although not easily. Making changes to the budget requires revisions to the SIP, which is a complicated and lengthy process that can make it more complicated for the MPO to start projects. If a budget revision is undertaken, MPOs should participate in the SIP revision process.

Although transportation conformity usually applies to road projects, there is a similar concept referred to as "general conformity." The general conformity requirements are similar to those of transportation conformity, but they apply to nonroad construction projects (40 CFR 93B [2002]; 40 CFR 51W [1999]). The primary difference between general and transportation conformity is the means by which conformity is shown. Unlike transportation conformity, for which there is generally a specific line item in the SIP for motor-vehicle emission budgets, general conformity is typically achieved by showing that emissions are kept below certain concentrations (40 CFR §§ 51.853(d), (i) [2003]; 93.153(b), (i) [2003]).

The air quality standards against which compliance is measured are "primary" standards. The CAA authorizes establishment of primary standards to protect public health and secondary standards to protect the environment. To date, EPA has not set any secondary standards at concentrations that are different from the primary standards. Although there is some concern with deposition of air pollutants into water bodies and other receiving environments that could warrant establishment of secondary standards, these effects are not directly regulated under the CAA. Under the current CAA scheme, ecological resources may be indirect beneficiaries of air pollution controls, but the program is directed at protection of public health.

## The Department of Transportation Act

Various specific provisions of the surface transportation laws and regulations of the Department of Transportation Act (23 USC § 101-158 [2003]) address environmental and resource protection. These are the environmental provisions of the Intermodal Surface Transportation Efficiency Act (ISTEA), the TEA-21, and earlier surface transportation statutes that govern FHWA activities.

### Parks and Public Areas

Section 4(f) of the Department of Transportation Act establishes certain criteria that must be met before a highway project can directly or indirectly use publicly owned parklands, wildlife protective areas, and other public recreational or historic sites (23 USC § 138 [2003]; 49 USC § 303 [2003]). FHWA regulations further explain these protections. The protected resources cannot be directly or indirectly affected without a specific finding that (1) no alternatives are feasible or prudent, and (2) the project includes all possible planning to minimize adverse effects (23 CFR § 771.135 [2002]). Section 4(f) can directly protect qualifying ecological habitat and can be an important aspect of the overall transportation-planning and wildlife-protection strategy.

## Native Vegetation

FHWA statutes and regulations encourage the use of native wildflowers in highway landscaping (see 23 USC. § 319(b) [2003]; 23 CFR 752 [2001]). However, only one-fourth of 1% of the funds expended on landscaping are required to be allocated to planting of native wildflowers on a project. Although the requirement is small, there is no limit on the opportunity to use landscaping to enhance native ecosystems.

## Clean Air Improvements

The Congestion Mitigation and Air Quality Improvement (CMAQ) Program (part of the TEA-21, Section 1110, 23 USC § 149 [2003]) provides additional air quality protection options. CMAQ applies to projects in areas that are not attaining the NAAQS or are trying to maintain attainment of the NAAQS. The sponsor of a CMAQ project must develop a proposal to improve air quality and submit it to the state and relevant MPO for approval. In concept, CMAQ projects could focus on ecological benefits if linked to air quality improvements.

This program provides flexible funding to local and state governments for transportation programs and projects to help meet the requirements of the CAA. Some of the eligible activities include

- Transit improvements.
- Travel demand management strategies.
- Traffic flow improvements.
- Public fleet conversions to cleaner fuels.

Additional funds are available for the areas that have not attained the NAAQS. Former nonattainment areas that have maintained compliance also have access to these funds. Funds are distributed to states based on a formula that considers an area's population by county and the severity of its air quality problems within the nonattainment or maintenance area. The CMAQ program emphasizes CO nonattainment and CO maintenance areas in relation to other priorities.

## Historic Bridges

The Historic Bridge Program (23 USC § 144(o) [2003]) provides that the reasonable costs associated with actions to preserve or reduce the impact of a project on the historic integrity of historic bridges is eligible as a reimbursable project cost. The program may provide some ecological benefits if historic bridges are located in association with ecological resources of concern, since the bridge protection would also protect those resources.

## Noise

DOT cannot approve any project on any federal-aid system unless it determines that the project includes adequate measures to implement the appropriate noise level standards (23 USC § 109(i) [2003]). The regulations promulgated require noise impact analysis, examination of potential mitigation measures, incorporation of reasonable and feasible noise abatement measures into the project, and coordination with local officials to provide helpful information on compatible land-use planning and control (23 CFR §§ 772.9, 772.11, 772.15 [2001]). The regulations contain noise abatement criteria that represent the upper limit of acceptable highway traffic noise for different types of land uses and human activities (23 CFR 772 [2001], Table 1). The regulations do not require that the abatement criteria be met in every instance. Rather, they require that every reasonable and feasible effort be made to provide noise mitigation when the criteria are approached or exceeded (23 CFR § 772.13(d) [2001]).

The intended beneficiary of the noise standards are humans, not wildlife. However, noise effects on wildlife are routinely considered under NEPA. Increasingly, noise from roads is evaluated as a direct or indirect impact on wildlife. Noise management features can protect wildlife habitat.

## Habitat Mitigation

TEA-21 specifies that federal funds may be used for "natural habitat and wetland mitigation efforts" related to federally funded projects (23 USC § 103(b)(6)(M) [2003]). These efforts may include participation in mitigation banks and contributions to state and regional conservation, restoration, enhancement, or creation. These mitigation projects can enhance the protection of ecosystems. Moreover, to the extent that mitigation banks allow the creation and restoration of continuous habitat, as opposed to segregated pockets of mitigation, the mitigation can be well situated in the landscape to meet multiple ecological functions.

## Context-Sensitive Design

Designers of road projects are admonished to take into consideration the environmental, scenic, aesthetic, historic, and community context of their roadway project (23 USC § 109 [2003]). So-called context-sensitive design (CSD) provides important opportunities to consider landscape and ecosystem functions in road planning. CSD can enhance the possibility of preservation for natural areas and, therefore, prevent the deterioration of ecosystem function. As addressed above, CSD principles are also emphasized in FHWA environmental policies.

## Transportation Enhancement Activities

Funds for transportation enhancement (TE) activities include the study and prevention of wildlife mortality. Funds may be used in all road systems except roads classified as local or rural minor collectors, unless these roads are on a federal-aid highway system. Federal transportation program funds allot 10% annually to each state for TE activities.

The FHWA field offices strongly encourage the state and MPOs to seek out and integrate TE activities into their plan development and programming processes. The metropolitan and statewide planning processes can occupy a central role in the identification, planning, and funding of TE activities. In particular, the planning processes are the appropriate mechanisms for determining funding priorities among competing TE activities, including those that are not part of specific transportation projects.

To be funded, TE activities must be included in the appropriate metro-politan and statewide transportation improvement activities. Given the widespread public interest in TE activities, they should be highlighted for public involvement implemented under the metropolitan and statewide requirements revised pursuant to TEA-21.

## General Environmental Policy

FHWA's environmental policies, addressed above, derive in part from specific directions in the transportation statutes. For example, economic, social, and environmental effects must be considered in planning and development of proposed projects on any federal-aid highway system over which FHWA exercises approval (23 USC 109(h) [2003]). Generally, given the similar scope and coverage, the NEPA documentation satisfies this statutory requirement.

Section 1309 of TEA-21 requires DOT to develop and implement a coordinated environmental review process for highway construction and mass transit projects. To that end, in July 1999, DOT, the Corps, DOI, EPA, NOAA-Fisheries, the U.S. Department of Agriculture (DOA), and the Advisory Council for Historic Preservation entered into a memorandum of understanding (MOU) under which they agreed to reduce project delays and protect and enhance environmental quality. The MOU identifies eight means of achieving the former and five ways of achieving the latter. These coordination mechanisms allow policy review of ecological concerns in an efficient manner.

## The Land and Water Conservation Fund Act

The Land and Water Conservation Fund Act (LWCFA) (16 USC §§ 4601(1)-4601(11) [2003]) is often used by state and local governments to obtain funds to acquire or improve parks and recreation areas. Section 6(f) of the LWCFA prohibits such property from being converted to another use without the approval of the National Park Service (NPS). This section also directs NPS to condition its approval of a conversion on the provision of replacement lands of equal value, location, and usefulness. Thus, replacement lands are required whenever conversions of Section 6(f) lands are proposed for highway projects. As with Section 4(f) of the transportation laws, described above, Section 6(f) provides funding for the direct protection of existing ecosystems.

## Historic and Archeological Preservation

Properties on or eligible for inclusion on the National Register of Historic Places are protected by Section 106 of the National Historic Preservation Act (NHPA) (16 USC § 470f [2003]). The law and regulations require coordination with the State Historic Preservation Officer and the Advisory Council on Historic Preservation with respect to qualifying properties. Historic preservation policies are also incorporated in Executive Order 11593 (see 23 CFR 771 [2001]; 36 CFR 60, 63 [2002]; 36 CFR 800 [2001]). The goals of the program are to avoid or mitigate damages to historic properties to the greatest extent possible.

Other laws provide similar protection for archeological resources and other special resources. For example, historic landmarks are protected under Section 110 of NHPA (16 USC § 470h-2 [2003]). Archeological resources must be recorded, surveyed, and preserved under the Archeological and Historic Preservation Act (16 USC §§ 469-469c [2003]), the Archeological Resources Protection Act (16 USC § 470aa-ii [2003]), and the Antiquities Act (16 USC §§ 431-433 [2003]). Native American resources are protected under the American Indian Religious Freedom Act (42 USC § 1996 [2003]), the Native American Grave Protection and Repatriation Act (25 USC § 3001 [2003]), and Executive Order 13007.

The preservation of historic areas is not intended to conserve ecological resources directly. However, to the extent that historic areas exist in conjunction with other protected areas or that historic areas are of a sufficient magnitude, their protection may enhance or provide an opportunity to enhance the protection of existing ecosystems.

## The Migratory Bird Treaty Act

The Migratory Bird Treaty Act (MBTA) (16 USC § 760e-760g [2003]) protects most common migratory wild birds in the United States from being killed, captured, collected, possessed, bought, sold, traded, shipped, imported, or exported. These birds are all common songbirds, waterfowl, shorebirds, hawks, owls, eagles, ravens, crows, native doves and pigeons, swifts, martins, swallows, and others. A complete list of protected species is found in Title 50, Section 10.13 of the *Code of Federal Regulations* (1989). The act also prohibits indirect killing of these birds by destroying their nests and eggs. However, such actions as road projects may be authorized to proceed under permits issued by the

USFWS pursuant to the regulations in Title 50, Parts 13 and 21 of the *Code of Federal Regulations* (1999). Federal-aid highway projects that are likely to result in the taking of birds protected under the act require a permit. Frequently, the permit will allow construction in bird habitat in accordance with such terms as seasonal limits to protect breeding periods or nest relocation programs.

The Migratory Bird Act directly protects covered wildlife, and as discussed with the ESA, the protection of such wildlife and their habitat often results in secondary benefits for otherwise unprotected species.

## The Farmland Protection Policy Act

The Farmland Protection Policy Act (FPPA) (7 USC §§ 4201-4209 [2003]) is intended to minimize farmland effects and maximize compatibility of federal actions with state and local farmland programs and policies. To achieve this goal, FHWA requires that NEPA documents (1) summarize the results of early consultation with the DOA's Natural Resources Conservation Service and, as appropriate, state and local agriculture agencies where agricultural land could be directly or indirectly affected by any alternative under consideration; (2) contain a map showing the location of all farmlands in the project impact area; (3) discuss the effects of the various alternatives; and (4) identify measures to avoid or reduce the effects. DOA rules regarding the FPPA are in Title 7, Part 658 of the *Code of Federal Regulations* (1987). The maintenance of existing farmland provides for a continuity of habitat for wildlife that uses such lands. For example, much of the migratory flyways in the central and western United States include farmlands that provide seasonal cover, forage, and breeding habitat for many species.

## The Coastal Zone Management Act

Coastal ecosystems are subject to the Coastal Zone Management Act (CZMA) (16 USC § 1451 [2003]) enacted to protect, preserve, develop, restore, and enhance coastal resources (16 USC § 1452 [2003]). States are eligible for federal funding if they opt to develop a coastal zone management plan, which is approved by NOAA. All federal agency activities affecting an area within the coastal zone must be carried out in a manner consistent with the relevant state's federally approved coastal zone management plan. The federal action agency must provide a "consistency determination" to the state coastal management agency at least 90 days before final approval of the federal activity, and the state agency must certify that the proposed activity complies with the enforceable policies of the state's approved program.

Among other things, coastal zone management plans must outline "areas of particular concern within the coastal zone" and describe how the management plan addresses and resolves the concerns for which the areas are designated (15 CFR § 923.21 [2003]). These are called "special area management plans," which provide an opportunity for detailing natural-resource protection of sensitive areas while allowing for reasonable economic growth. Additional regulations concerning implementation of the CZMA are in Title 15, Part 923 of the *Code of Federal Regulation* (2003). The CZMA directly protects coastal ecosystems by addressing the nature of development and the destruction of habitat.

## Federal Fisheries Conservation and Management Act (Magnuson Act)

Now known as the Magnuson Fishery Conservation and Management Act (16 USC § 1801 [2003]), this law was passed in 1976 to establish protection for off-shore and anadromous fish species. Congress recognized that many fisheries had been depleted from over fishing. It found that healthy fish stocks were important not only for ecological reasons but also for the viability of important commercial and recreational fishing industries. Congress expressly acknowledged the importance of habitat protection to the maintenance of healthy fish stocks.

Under the law, fishery management plans are developed by regional fishery management councils, establishing catch limits and other conditions to protect stocks from over fishing (16 USC §§ 1851-1860 [2003]). Fishery management plans developed under the law, particularly for anadromous species, such as salmonids as well as many commercial species, can and do include controls on land uses in or near fish habitat. As a result, for road projects that cross waterways used by certain fisheries, this program will have additional requirements for ecological protection of fish habitats and consultation with pertinent federal agencies.

## Executive Orders

Federal agencies must consider various environmental and resource issues under direction of an Executive Order (EO). In most instances, the agency will try to incorporate compliance with the EOs in either the NEPA documentation or in its formal record of decision.

EO 13274 (2002), addressed previously, establishes policies for transportation infrastructure reviews and environmental stewardship. It states that executive departments and agencies are to take appropriate actions to the extent consistent with the law to expedite reviews of high-priority transportation projects and promote environmental stewardship in the nation's transportation system (67 Fed. Reg. 59449 [2002]). It requires the adoption of policies and procedures necessary to accomplish that goal and establishes a task force for assembling a list of high-priority projects and monitoring compliance with the EO. Similar goals with regard to environmental streamlining were established in TEA-21, as discussed above.

EO 12898, Environmental Justice (1994), requires consideration of the environmental and social effects of an action on minority and low-income populations to ensure that such populations are not disproportionately affected. To implement these policies, DOT issued a Final Environmental Justice Strategy (DOT Order 5610.2) dated April 15, 1997. FHWA issued an order (DOT Order 6640.23) dated December 2, 1998, describing the actions necessary to protect environmental justice concerns (FHWA 2002c). Natural systems can be indirect beneficiaries of this evaluation of environmental effects at the neighborhood or community scale.

EO 11988, Floodplain Management, as amended by EO 12148, is intended to avoid the short- and long-term adverse effects associated with the development of floodplains. It requires the assessment of floodplain hazards and specific findings in the final environmental document for significant floodplain encroachments. As with the Section 404 program, protection of land within floodplains enhances habitat directly through preservation and indirectly through improvements in water quality. FHWA provides various policy guidance on implementation of this EO (DOT Order 5650.2) (23 CFR § 650, 771 [2004]) (FWHA 2004 a,b).

EO 13112, Invasive Species (1999), is directed at the control of invasive alien species (64 Fed. Reg. 6183 [1999]) (National Invasive Species Council 2004). Such species have the potential to overrun native species and transform ecological systems. The EO prevents agencies from authorizing, funding, or carrying out actions that are likely to cause or promote the introduction or spread of invasive species. FHWA generally complies with the mandate of the EO through the NEPA process. The EO protects the loss of native species at the hands of exotics, clearly preserving existing ecosystems. FHWA has developed guidance to encourage environmentally beneficial landscape practices.

EO 13186 supplements the regulations under the Migratory Bird Treaty Act (66 Fed. Reg. 3853 [2001]). It directs agencies to enter into MOUs with the USFWS to establish procedures to promote conservation of bird populations (Shrouds 2001). The order does not apply to federal-aid highway projects that are carried out by nonfederal entities, such as state DOTs.

EO 11990 (1977) also addresses protection of wetlands. This EO states that federal agencies should avoid direct or indirect support of activities adversely affecting wetlands where there is an alternative. The NEPA documentation or the Section 404 Record of Decision generally records compliance with EO 11990.

## OTHER TYPES OF LAWS

### Various Federal Resource Laws

In addition to the topically oriented statutes addressed above, Congress has enacted many statutes concerning specific kinds of ecological resources that could, under some circumstances, relate to roads. Because these laws may only occasionally involve road projects, they are not addressed in any detail in this chapter. For example, road projects in Florida could involve compliance with the Marine Mammal Protection Act (16 USC §§ 1361-1421h [2003]) in relation to manatees. The Bald and Golden Eagle Act (16 USC § 668 [2003]) prohibits destruction of bald and golden eagles or their nests. National forests are subject to an extensive system of management administered largely by the DOA Forest Service. Not only forest roads but other road projects affecting national forests would involve attention to this body of law.

Other statutes and programs provide funding and policy for specified ecological systems, such as estuarine restoration (33 USC §§ 2901-2909 [2003]). EPA also administers a national estuarine program using authorities under the CAA (EPA 2003b). These kinds of research and financial assistance programs provide sources of information or advocacy for particular ecological resources but rarely establish requirements that road projects must meet. A statute such as the Wild and Scenic Rivers Act (16 USC §§ 1271-1287 [2003]) affords states the opportunity to designate rivers as wild or scenic and obtain additional protection of the resource from degradation. Some road projects may involve these kinds of federally protected or managed resources as well. Other similar examples include statutes such as the Wild Horses and Burros Act (16 USC §§ 1331-1340 [2003]), which protects wild free-roaming horses and burros.

Different organizations might include different laws in a compilation of environmental laws related to transportation. In addition to the laws as addressed in this chapter, FHWA (2004a) identified laws pertinent to environmental compliance. The Defenders of Wildlife (2004) assembled a list of laws related to wildlife and transportation. The laws largely correspond to the laws addressed in this chapter. The Defenders of Wildlife added some statutes, such as the Coastal Barrier Resources Act and the National Trails Systems Act. The Transportation Research Board analysis (Blaesser et al. 2003) uses a slightly different list, having greater consideration for possible contamination and liability. Although different entities may include slightly different lists of laws related to resources or wildlife in discussing transportation projects, this chapter addresses the major laws involved.

## State Laws

Although this chapter does not attempt to summarize any nonfederal environmental laws, state and local laws play a major role in addressing ecological concerns associated with roads. Land-use planning, managed by local and regional authorities, substantially affects ecological systems and functions. This planning addresses not only growth (such as zoning for commercial, industrial, and residential development) but also siting of parks, recreation areas, and natural preserves. Local laws run the full gamut of environmental and ecological issues. Some states have established protections for specific ecosystem types, such as coastal or riparian zones, by restricting construction within a certain distance of water resources. The nature of local laws reflects community priorities and needs. States establish and enforce the "best management practices" to protect waterways from siltation and to prevent erosion effects of construction. There are many other examples of types of state and local laws.

For several major federal statutes, notably the Clean Water Act and the CAA, the federal standards will be implemented most frequently by states and localities through delegated programs. As a practical matter, therefore, the road project will seek compliance with these laws by obtaining a state water permit and a state air-emission permit. State programs that are more stringent than federal programs will include requirements of federal law plus any additional state requirements.

At the transportation planning stage, which occurs largely at the regional and local level, state and local ecological laws could play a major role. Project development and implementation is also subject to all applicable state and local laws. Thus, there are opportunities at nonfederal levels to establish standards for evaluation of ecological concerns for transportation planning and project development.

## DISCUSSION OF LAWS AND ECOLOGICAL CONCERNS

To assess whether and how the law addresses "ecological concerns," a working definition of ecological concerns, is needed. For purposes of this analysis, the committee determined that ecological concerns should include species or habitats of some recognized value in a range of landscapes. Not every landscape is of ecological concern, and the assignment of ecological functions and values may be subject to a combination of scientific and social processes. Ecological concerns arise at many scales, from genetic to species or populations to ecosystems.

Separately, the committee recognized that ecological concerns also encompass a cross-disciplinary spatial method of viewing landscapes and changes to them. An ecological approach may provide a useful way to capture interrelated or cumulative effects of actions within a pertinent landscape sphere. Spatially, this ecological concern addresses the broader perspective of looking at how various actions may together have particular consequences.

Using the broad concepts of species and habitats and spatial landscapes as ecological concerns, the committee considered the ways that existing laws address the following:

- Terms requiring consideration of any ecological concerns.
- Terms protecting species or habitats of recognized value.
- The stage at which ecological concerns are considered.
- The weight given to ecological concerns.

Notably, the ecological concerns at regional, multistate, or national levels are largely unaddressed in the current legal structure. The transportation planning process, which is a regional process, is subject to almost no mandatory provisions with respect to ecological concerns. Under federal law, LRTPs and TIPs can be approved by the MPO and reviewed by FHWA without consideration of ecological effects. FHWA's general policies encourage consideration of environmental issues during long-range planning, but the requirements are not enforceable. The only environmental concern that must be satisfied during transportation planning is demonstration of CAA conformity. At ecological scales beyond an MPO jurisdiction, which may be regional, state, or national, there are no legal requirements to consider transportation and ecological concerns in tandem.

The planning stage offers an excellent opportunity to incorporate multiple kinds of ecological concerns to ensure that regional ecological resources are addressed well before specific project planning. For ecological resources that extend beyond the jurisdiction of one MPO, coordination of transportation planning and ecological concerns can address the ecological component early in the process.

At project development and implementation, many environmental laws apply. The laws summarized within this chapter address ecological concerns in various ways. For purposes of discussion, the committee classified the laws in terms of several kinds of ecological issues, as shown in Tables 5-1 through 5-5.

## Legal Base of Environmental Policy

Table 5-1 organizes the federal laws in terms of their consideration of ecological system components that are pertinent to ecologists. For each law, the table indicates what kind of terms the statute provides. "Mandatory" (M) reflects a legal provision that provides some type of third-party enforceable protection of the ecosystem component, usually by requiring a permit for disturbance. "Discretionary" (D) indicates that the law requires disclosure and discussion of the issue but does not mandate a particular outcome from that discussion. D also means that the issue can be addressed if the lead agency considers it pertinent or relevant. "Not considered" (NC) means that the law is silent or the component is not pertinent to the law. The committee recognizes that assigning these designations in the tables involves potential error from oversimplification but believes that, as a general rule, the legislation has been accurately classified.

As Table 5-1 demonstrates, relatively few federal programs impose mandatory, enforceable protection of ecosystem components. Many laws allow or compel consideration of ecological concerns but are discretionary. Consideration of ecological concerns allows for public notice of ecological issues and provides data on which agency actions can be modified. As a matter of NEPA policy, FHWA requires mitigation for all environmental effects identified in an EIS. For FHWA-funded projects, NEPA mitigation commitments in a record of decision are enforceable terms of the federal funding agreement. These terms are thus enforceable between FHWA and a state but may not be enforceable by outside third parties. As a result, although NEPA itself only requires consideration of ecological concerns, additional protection is provided through administration of highway funding laws. For most laws, as summarized in Table 5-1, a discretionary factor can be balanced against other factors, providing flexibility but also, arguably, giving less weight to individual discretionary factors.

Table 5-2 organizes the legislation in terms of certain types of species addressed or covered by the programs, using the same three categories: mandatory, discretionary, or not considered. Using these categories is another way to evaluate whether and how the laws consider issues of importance to ecologists. Relatively few laws address particular types of species. Only the ESA, the MBTA, and the Section 404 program provide mandatory terms, requiring permits or other authorizations, for selected species. Clearly, the vast majority of species are not addressed by legislation, process, or executive order.

Table 5-3 orders legislation according to ecosystem type, asking whether and how the regulations address certain types of ecosystems. This table also uses the same three categories: mandatory, discretionary, or not considered. A number of federal laws are designed to address particular ecosystems and provide for permits or other clearances for activities within those ecosystems. In other words, Congress incorporated the concept of ecosystem types in some legislation and designated certain types of ecosystems for special protection. Many laws authorize the regulators to view issues in terms of ecosystems. However, as with species, only a few of the many types of ecosystems are addressed through legislation.

Table 5-4 summarizes the ecological scale at which the program applies. For this table, the summary indicates whether the law requires

**TABLE 5-1** Types of Considerations Required by Environmental Laws, Programs, and Executive Orders That Address Ecosystems and Components of Ecosystems with Regard to Roads[a]

| Legislation or Program | Component of Ecological System | | | | | |
| --- | --- | --- | --- | --- | --- | --- |
| | Physical | | | Biotic | | |
| | Water | Soil | Other[b] | Genetic | Species/ Populations | Ecosystem |
| MPO planning process | D | D | D | NC | D | D |
| FHWA environmental policies | M | M | D | NC | D | M |
| NEPA—In general | D | D | D | D | D | D |
| EIS | M | M | M | NC | D | D |
| EA | M | D | D | NC | D | D |
| Categorical Exclusion | NC | NC | NC | NC | NC | NC |
| Clean Water Act | | | | | | |
| Section 402 NPDES program | M | NC | | NC | D | D |
| Section 404 dredge and fill permits | M | D | D | D | D | D |
| Nonpoint-source pollution | D | D | NC | NC | NC | D |
| Rivers and Harbors Act | M | D | NC | NC | D | D |
| Endangered Species Act | D | D | D | D | M | D |
| Clean Air Act | | | | | | |
| When setting standards | NC | NC | M | D | D | D |
| For road projects | NC | NC | M | NC | NC | NC |
| Department of Transportation Act | | | | | | |
| Parks and public areas | M | M | M | NC | NC | M |

|  |  |  |  |  |  |  |
|---|---|---|---|---|---|---|
| Native vegetation | M | M | M | NC | NC | D |
| CMAQ | NC | NC | M | NC | NC | D |
| Historic bridges | D | NC | M | NC | NC | D |
| Noise | NC | NC | M | D | D | D |
| Habitat mitigation | D | D | D | D | D | D |
| Context-sensitive design | D | D | D | NC | D | D |
| Transportation enhancements | D | D | D | NC | D | D |
| Historic Preservation Act | NC | NC | M | NC | NC | NC |
| Land and Water Conservation Fund | NC | NC | NC | NC | NC | NC |
| Migratory Bird Treaty Act | NC | NC |  | D | M | D |
| Farmland Protection Policy Act | D | D | M | NC | NC | NC |
| Coastal Zone Management Act | M | D | M | D | D | M |
| Magnuson Fishery Act | D | NC |  | M | M | D |
| Executive Orders |  |  |  |  |  |  |
| EO 13274 environmental stewardship | D | D | D | D | D | D |
| EO 12898 environmental justice | NC | NC | M | NC | NC | D |
| EO 11988 floodplains | D | D | M | NC | D | D |
| EO 13112 invasive species | D | D | M | NC | D | D |
| EO 11990 wetlands | M | M | M | NC | D | M |
| EO 13186 migratory birds | D | NC | M | NC | M | D |

aComponents in table describe whether the law requires the following: consideration of the component is mandatory (M), consideration of the component is discretionary (D), or considerations of the component is not required (NC).
bIndicates that the law may involve other physical matters, such as geological or geophysical issues

**TABLE 5-2** Types of Considerations Required by Environmental Laws, Programs, and Executive Orders That Address Groups of Organisms (Species and Populations) with Regard to Roads[a]

| Legislation or Program | Invertebrates | Migratory Birds | Migratory Fish | Endangered Species | Amphibians | Plants |
|---|---|---|---|---|---|---|
| MPO planning process | NC | D | D | D | D | D |
| FHWA environmental policies | D | D | D | M | D | D |
| NEPA—in general | D | D | D | M | D | D |
| EIS | D | M | D | M | D | D |
| EA | D | M | D | M | D | D |
| Categorical Exclusion | NC | NC | NC | NC | NC | NC |
| Clean Water Act | | | | | | |
| Section 402 NPDES program | D | NC | D | M | D | D |
| Section 404 dredge and fill permits | NC | M | M | M | D | M |
| Nonpoint-source pollution | NC | M | D | M | D | D |
| Rivers and Harbors Act | NC | NC | D | M | D | M |
| Endangered Species Act | M | M | M | M | M | M |
| Clean Air Act | | | | | | |
| When setting standards | NC | NC | NC | NC | NC | NC |
| For road projects | NC | NC | NC | D | NC | D |
| Department of Transportation Act | | | | | | |
| Parks and public areas | NC | NC | NC | NC | NC | NC |
| Native vegetation | NC | NC | NC | NC | NC | M |

| Component | | | | | | |
|---|---|---|---|---|---|---|
| CMAQ | NC | NC | NC | NC | NC | NC |
| Historic bridges | D | NC | M | D | NC | NC |
| Noise | NC | NC | D | NC | D | NC |
| Habitat mitigation | D | D | M | D | D | D |
| Context-sensitive design | D | D | M | D | D | D |
| Transportation enhancements | D | D | D | D | D | D |
| Historic Preservation Act | NC | NC | NC | NC | NC | NC |
| Land and Water Conservation Fund | NC | NC | NC | NC | NC | NC |
| Migratory Bird Treaty Act | NC | NC | M | NC | M | NC |
| Farmland Protection Policy Act | NC | NC | NC | NC | NC | NC |
| Coastal Zone Management Act | D | D | M | D | D | D |
| Magnuson Fishery Act | NC | D | M | M | NC | NC |
| Executive Orders | | | | | | |
| EO 13274 environmental stewardship | D | D | D | D | D | D |
| EO 12898 environmental justice | NC | NC | NC | NC | NC | NC |
| EO 11988 floodplains | NC | NC | NC | NC | NC | NC |
| EO 13112 invasive species | M | NC | NC | NC | NC | NC |
| EO 11990 wetlands | D | D | M | M | D | D |
| EO 13186 migratory birds | NC | D | M | NC | M | NC |

aComponents in table describe whether the law requires the following: consideration of the component is mandatory (M), consideration of the component is discretionary (D), or consideration of the component is not required (NC).

**TABLE 5-3** Types of Considerations Required by Environmental Laws, Programs, and Executive Orders That Address Different Ecosystem Types with Regard to Roads[a]

| Legislation or Process | Wetlands | Freshwater | Coastal Marine | Forests | Meadows/Prairies |
|---|---|---|---|---|---|
| MPO planning process | D | D | D | D | D |
| FHWA environmental policies | D | | D | D | D |
| NEPA—in general | D | D | D | D | D |
| EIS | D | D | D | D | D |
| EA | D | D | D | D | D |
| Categorical Exclusion | NC | NC | NC | | NC |
| Clean Water Act | | | | | |
| Section 402 NPDES program | M | M | M | D | D |
| Section 404 dredge and fill permits | M | M | M | M | M |
| Nonpoint-source pollution | D | D | D | D | D |
| Rivers and Harbors Act | M | M | M | NC | NC |
| Endangered Species Act | M | M | M | M | M |
| Clean Air Act | | | | | |
| When setting standards | NC | NC | NC | NC | NC |
| For road projects | NC | NC | NC | NC | NC |
| Department of Transportation Act | | | | | |
| Parks and public areas | D | D | D | D | D |
| Native vegetation | M | M | M | M | M |
| CMAQ | NC | NC | NC | NC | NC |

| | | | | | |
|---|---|---|---|---|---|
| Historic bridges | D | D | D | D | D |
| Noise | NC | NC | NC | NC | NC |
| Habitat mitigation | D | D | D | D | D |
| Context-sensitive design | D | D | D | D | D |
| Transportation enhancements | D | D | D | D | D |
| Historic Preservation Act | NC | NC | NC | NC | NC |
| Land and Water Conservation Fund | NC | NC | NC | NC | NC |
| Migratory Bird Treaty Act | D | D | D | D | D |
| Farmland Protection Policy Act | D | NC | NC | NC | D |
| Coastal Zone Management Act | D | M | M | D | D |
| Magnuson Fishery Act | D | M | M | NC | NC |
| Executive Orders | | | | | |
| EO 13274 environmental stewardship | D | D | D | D | D |
| EO 12898 environmental justice | NC | NC | NC | NC | NC |
| EO 11988 floodplains | M | M | M | D | D |
| EO 13112 invasive species | D | D | D | D | D |
| EO 11990 wetlands | M | M | M | D | D |
| EO 13186 migratory birds | D | D | D | D | D |

*a* Entries in table describe whether the law requires the following: consideration of the component is mandatory (M), consideration of the component is discretionary (D), or consideration of the component is not required (NC).

**TABLE 5-4** Ecological Scale at Which Legislation Is Applied[a]

| Legislation or Process | Scale Unit | | | | |
|---|---|---|---|---|---|
| | Road Segment | Landscape | Watershed | Eco-region | Nation |
| MPO planning process | Assess | Assess | Assess | Assess | NC |
| FHWA environmental policies | Assess | Assess | Assess | Assess | NC |
| NEPA—in general | Assess | Assess | Assess | Assess | NC |
| EIS | Assess | Assess | Assess | Assess | NC |
| EA | Assess | Assess | Assess | NC | NC |
| Categorical Exclusion | NC | NC | NC | NC | NC |
| Clean Water Act | | | | | |
| Section 402 NPDES program | Limit | Limit | Assess | NC | NC |
| Section 404 dredge and fill permits | Limit | Limit | Assess | Assess | NC |
| Nonpoint-source pollution | Limit | Assess | Assess | Assess | NC |
| Rivers and Harbors Act | Limit | Limit | Assess | NC | NC |
| Endangered Species Act | Limit | Assess | Assess | Assess | NC |
| Clean Air Act | | | | | |
| When setting standards | NC | NC | NC | NC | NC · |
| For road projects | Assess | Assess | NC | NC | NC |
| Department of Transportation Act | | | | | |
| Parks and public areas | Limit | Limit | NC | NC | NC |
| Native vegetation | Limit | Limit | NC | NC | NC |

| | | | | | |
|---|---|---|---|---|---|
| CMAQ | NC | NC | NC | NC | NC |
| Historic bridges | NC | NC | NC | NC | NC |
| Noise | NC | Assess | NC | NC | NC |
| Habitat mitigation | Assess | Assess | Assess | Assess | Assess |
| Context-sensitive design | Assess | Assess | Assess | Assess | Assess |
| Transportation enhancements | Assess | Assess | Assess | Assess | NC |
| Historic Preservation Act | Assess | Assess | NC | NC | NC |
| Land and Water Conservation Fund | Limit | Limit | NC | NC | NC |
| Migratory Bird Treaty Act | | Limit | Assess | Assess | Assess |
| Farmland Protection Policy Act | Assess | Assess | NC | NC | NC |
| Coastal Zone Management Act | Limit | Limit | Assess | Assess | NC |
| Magnuson Fishery Act | Limit | Limit | Assess | Assess | Assess |
| Executive Orders | | | | | |
| EO 13274 environmental stewardship | Assess | Assess | Assess | Assess | NC |
| EO 12898 environmental justice | NC | NC | NC | NC | NC |
| EO 11988 floodplains | Assess | Assess | Assess | Assess | NC |
| EO 13112 invasive species | Limit | Limit | NC | NC | NC |
| EO 11990 wetlands | Limit | Limit | NC | NC | NC |
| EO 13186 migratory birds | Limit | Limit | NC | NC | NC |

*a*The table indicates whether the law requires an assessment at a particular scale (Assess); the law establishes some protection, such as a permit requirement or other limitation (Limit); or scale is not considered (NC).

an assessment at a particular scale (Assess); the law establishes some protection, such as a permit requirement or other limitation (Limit); or scale is not considered (NC). This table illustrates that most laws are targeted at a local scale and are related to a particular project.

In conjunction with specific projects, some laws require consideration of indirect, cumulative, or other related effects, which are reflected in a designation that the law requires assessment of larger ecological scales.

Table 5-5 addresses the political scale at which the legislation is applied. Although all the laws are federal, the table indicates what political levels are pertinent under each statute in terms of application and compliance with the law. The level of government that bears primary responsibility for compliance is identified.

## Environmental Reviews and Roads

Information on the level of environmental review of road projects under NEPA is sparse. One study by the American Association of State Highway and Transportation Officials (AASHTO) surveyed 32 states and found that 92% of environmental documents processed by these state DOTs were categorical exclusions (CE) (STPP 2002, TransTech Management, Inc. 2000). Environmental assessments comprised 7% of the documents, and full EIS were covered in 2% of the cases.

As stated in Chapter 2, few studies have attempted to assess how the legal requirements of NEPA and other environmental statutes have affected the length of time needed to complete road projects (GAO 1994, FHWA 2003f). These studies used information and data from sets of past highway projects for which NEPA EIS or Section 404 reviews were implemented. According to GAO (1994), these projects represented a small percentage (less than 3%) of all road projects. In this small set of projects, however, required environmental reviews increased the time of project completion.

The time associated with environmental review varied as a function of the type of review (CE, finding of no significant impact [FONSI], EA, or EIS). In 1994, GAO examined 76 road projects completed from 1988 to 1993 to determine how an EIS extended the time to project completion. It found that those projects took from 2 to 8 years to complete the necessary NEPA and Section 404 reviews. An EIS prepared for NEPA review averaged 4.4 years, and an EIS under both NEPA and Section 404 permits averaged 5.6 years to complete.

In contrast to the EIS findings, FHWA assessed projects that involved either a FONSI or a CE. Rather than using project data, such as start and completion date, this assessment was done by polling division offices. The agency concluded that most FONSI reviews took about 18 months to complete, and CE reviews took about 6 months to complete. AASHTO (TransTech 2000) did a survey of state DOTs and reported that about half of CEs were generally delayed and actually took between 1 and 2 years to complete.

FHWA and the Berger group (FHWA 2003f) attempted the most quantitative analysis of this issue by calculating statistics from about 100 road projects done between 1970 and 2000. All of these projects required an EIS pursuant to NEPA. This group concluded that the mean time to complete an EIS was about 3.6 years, which accounted for about 28% of the total project development process. They also noted that the time to complete an EIS doubled from the 1970s (about 2.2 years) to the 1990s (about 5 years). This report and others suggest that the current data collected on environmental review of road projects raise questions about the efficacy of NEPA, since the lengthy NEPA process can lead to frustration on all sides. The report's authors concluded by suggesting that their study provides a useful baseline against which streamlining efforts could be evaluated.

## ISSUES

### Issues and Gaps Related to Ecology

A number of gaps exist within the array of environmental legislation, regulations, and policies. Not all species or ecosystem types obtain protection from adverse effects associated with roads. By definition, some of these acts require addressing only parts of ecological systems, such as wetlands, endangered species, and clean water. Little protection is afforded to species or habitats that are not rare, threatened, or endangered, even if those habitats serve important ecological functions. For example, an ecosystem patch or zone may perform functions of flood control, groundwater recharge, wildlife habitat, or open-space aesthetics. Without special designation under federal law, these ecosystem functions escape protection.

**TABLE 5-5** Political Scale at Which Environmental Legislation, Program, or Executive Orders Are Applied[a]

| Legislation or Program | Local | County | Multicounty | State | Federal |
|---|---|---|---|---|---|
| MPO planning process | Limit | Limit | NC[b] | NC | NC |
| FHWA environmental policies | Assess | Assess | Assess | Assess | Assess |
| NEPA—in general | Assess | Assess | Assess | Assess | Assess |
|   EIS | Assess | Assess | Assess | NC | NC |
|   EA | Assess | Assess | Assess | NC | NC |
|   Categorical Exclusion | NC | NC | NC | NC | NC |
| Clean Water Act | | | | | |
|   Section 402 NPDES program | Limit | NC | NC | NC | NC |
|   Section 404 dredge and fill permits | Limit | NC | NC | NC | NC |
|   Nonpoint-source pollution | Assess | Assess | Assess | NC | NC |
| Rivers and Harbors Act | Limit | NC | NC | NC | NC |
| Endangered Species Act | Limit | Assess | Assess | NC | NC |
| Clean Air Act | | | | | |
|   When setting standards | Limit | Assess | Assess | NC | NC |
|   For road projects | Limit | Assess | Assess | NC | NC |
| Department of Transportation Act | | | | | |
|   Parks and public areas | Limit | NC | NC | NC | NC |
|   Native vegetation | Limit | Limit | Limit | NC | NC |
|   CMAQ | Assess | Assess | Assess | NC | NC |

| | | | | | |
|---|---|---|---|---|---|
| Historic bridges | Limit | NC | NC | NC | NC |
| Noise | Limit | NC | NC | NC | NC |
| Habitat mitigation | Assess | Assess | Assess | NC | NC |
| Context-sensitive design | Assess | Assess | Assess | NC | NC |
| Transportation enhancements | Assess | Assess | Assess | NC | NC |
| Historic Preservation Act | Limit | NC | NC | NC | NC |
| Land and Water Conservation Fund | Limit | NC | NC | NC | NC |
| Migratory Bird Treaty Act | Limit | Assess | Assess | Assess | NC |
| Farmland Protection Policy Act | Assess | Assess | Assess | Assess | NC |
| Coastal Zone Management Act | Limit | Limit | Limit | NC | NC |
| Magnuson Fishery Act | Limit | Assess | Assess | NC | NC |
| Executive Orders | | | | | |
| EO 13274 environmental stewardship | Assess | Assess | Assess | Assess | Assess |
| EO 12898 environmental justice | Limit | Limit | NC | NC | NC |
| EO 11988 floodplains | Assess | Assess | Assess | NC | NC |
| EO 13112 invasive species | Limit | Limit | Limit | Assess | NC |
| EO 11990 wetlands | Limit | Assess | Assess | NC | NC |
| EO 13186 migratory birds | Limit | Limit | Assess | Assess | Assess |

[a]The table indicates whether the law requires an assessment at a particular scale (Assess); the law establishes some protection, such as a permit requirement or other limitation (Limit); or scale is not considered (NC).

[b]MPOs may be multicounty, in which case this entry could be Limit.

Another gap is a knowledge gap in our ability to predict environmental effects, as required under NEPA and other acts. One of NEPA's fundamental assumptions is that effects—direct, indirect, or cumulative—can be predicted. For at least 20 years, there has been a broad body of ecological literature that suggests that the ability to assess these effects is variable, ranging from fairly well-understood phenomena to inherently unpredictable systems. For example, most hydrological dynamics are understood well enough within streams and rivers to know how to minimize hydrological effects associated with bridges. Other complex biological phenomena, such as broad-scale migrations, patterns of exotic biota invasions, and population dynamics, are often difficult to predict and assess impacts prior to action. The array of laws and orders that underpin assessment and planning actions are usually only applied at local or small scales, the bounds determined for political or social reasons, not ecological ones. The legislative guidance (Table 5-5) suggests that environmental reviews can be conducted at larger spatial scales (state and national level) but are generally not done at these larger scales. This finding corresponds to the scales described in Chapter 3, showing that most of our knowledge of impacts associated with roads occurs at small spatial and temporal scales. The reasons for choosing smaller scales for environmental review are unclear but discussed further in Chapter 6.

Other gaps exist in the laws and orders that relate to environmental evaluations. No legal requirements were found that compel assessment of road effects on ecosystem goods and services. It is possible that some of the knowledge gaps identified in Chapter 3, especially road effects on ecosystem goods and services, are due in part to no legal requirement.

Existing laws are limited in capturing cumulative, indirect, and synergist effects of road systems. Although cumulative and indirect effects must be included under NEPA, the analysis generally does not consider long-term cumulative effects of an entire transportation system because NEPA applies at the project development scale. The MPO is not required to evaluate cumulative or indirect ecological effects of the LRTP or TIP.

## Issues of Multiple Agency Involvement

The participation of multiple agencies with responsibility for ecological and transportation concerns raises other issues. As reflected in the statutes described in this chapter, multiple layers of coordination and consultation are the rule, not the exception, in uniting transportation and ecological concerns. The resource agencies that are responsible for ecological matters may lack sufficient personnel, funding, or knowledge of transportation to perform their roles smoothly. In transportation, there are substantial differences in resource staffing and expertise among local entities, MPOs, and state DOTs. Local entities and MPOs rarely have sufficient budgets to maintain ecological expertise on the staff. In contrast, state DOTs have considerable ecological expertise. Coordination of all of these multiple agency resources can be difficult at the planning stage and the project stage.

Not only are the federal environmental laws administered by agencies other than FHWA or the state DOTs, but also there is no single agency with paramount authority for ecological resources. DOT has final authority for federal transportation projects, but the projects still require multiple agency review and permission. As described, the laws addressed in this chapter involve EPA, USFWS, the Corps, NPS, and other federal resource agencies. These resource agencies have multiple duties in addition to those related to transportation and generally have small staffs to work on transportation issues. The resource agencies often have pertinent ecological information, but it might not be maintained in easily accessible databases. Involvement of multiple federal agencies requires coordination of personnel time and work priorities. In some instances, state DOTs have provided a resource agency with funding to make personnel available to ensure attention to transportation projects.

The resource agencies are generally structured on a different regional basis than FHWA or the states. For example, EPA operates through 10 regional offices, the Corps has 35 district offices, and the USFWS has 8 regional offices, although there are also field offices with personnel from the ecological service branches in most states. These structural differences can make interagency coordination challenging at best.

There also can be conflicts among agencies involved with transportation projects. Disputes between a state DOT and EPA or USFWS over ecological effects of road projects are not uncommon. Having so many agencies involved in ecological issues offers the advantage of detached independent analysis of issues but often at the price of delays or controversies. Executive Order 13274, Environmental Stewardship and Transportation Infrastructure Project Reviews (September 2002), is designed to address potential interagency issues by establishing a process for review of projects nominated by states for special attention.

## Issues Concerning Programmatic, Proactive Approaches

Ecological concerns may extend beyond the spatial and temporal boundaries of specific projects. Some federal laws offer opportunities to look at environmental concerns programmatically, encompassing a broad geographic or temporal area. At a programmatic rather than a specific project level, a broader range of ecological concerns can be addressed.

For example, under NEPA, agencies are authorized to conduct a programmatic environmental impact statement, looking at the environmental effects of more than one project. A "program" can be defined by the agency. FHWA could consider a series of road projects that are planned for a region as a "program" under NEPA. The programmatic review may assist early planning to integrate ecological concerns into long-term planning. Thus, a programmatic review may identify wildlife or other sensitive environments so that individual projects can be planned accordingly to avoid and protect such ecosystems.

Other laws, such as ESA, do not provide for any programmatic or abbreviated analysis. The statute does not contemplate programmatic ESA review for territories or regions where multiple effects might be proposed from diverse sources. Although FHWA or the state DOT could request ESA review for multiple effects in a single location, they would have to have fairly specific information about the various projects, as well as the species of concern. Generally, at the LRTP or TIP stage, there is not sufficient specific information about particular projects to allow a meaningful review for effects on threatened or endangered species. If the law either required or provided a mechanism to consider threatened or endangered species regionally in advance of specific transportation projects, protection of certain wildlife and habitat could be assisted at the early planning stages.

The federal laws were written with a focus on environmental effects of specific projects. Despite some options to apply the laws in a more programmatic, broader fashion, limitations on resources impose practical constraints on such efforts.

## SUMMARY

Transportation projects are primarily developed from the "bottom up," reflecting community needs within local political jurisdictions. There are few incentives or disincentives in law for communities to consider effects beyond the political jurisdictions, including ecological effects. The decision making has been primarily local, but the ecological effects can extend beyond the local area. Federal transportation planning law relies heavily on local jurisdictions, thus creating few opportunities for evaluation of ecosystem concerns, especially at broader scales.

The transportation planning system, carried out by MPOs, offers an opportunity for consideration of ecological concerns at an early stage in road planning. Incorporating ecological concerns at the planning stage would assist in providing protection for ecological functions and in avoiding controversies at the project development and implementation stage. Only NEPA considers effects to the environment at other scales, including indirect or cumulative effects. However, NEPA does not apply to the MPO planning processes.

Many entities interested in ecological concerns, including resource agencies, nongovernmental entities, and the public, do not understand the relationship of federal environmental laws to the transportation planning process or the project development and implementation process. This lack of understanding can result in controversies over specific projects after years of planning by transportation entities.

Although a wide range of laws, regulations, and policies require some degree of consideration of ecological effects of road construction, the existing legal structure leaves significant gaps. Road projects need only permits to conduct activities that might have an impact on certain types of ecological features—wetlands, endangered species, or migratory birds—and generally at a small scale. Moreover, the permit programs usually consider only direct effects of a road (construction or use) on the protected resource.

The data necessary to evaluate efficacy of policy and law are not easily accessible or amenable to synthesis. The data are contained in project-specific EISs, EAs, records of decision, or permits (for example, wetlands permits) that are not easily available for research.

With few exceptions (for example, wetlands and endangered species), existing law authorizes ecological concerns to be balanced with goals of transportation mobility, capacity, and other social needs in determining whether and how to undertake transportation projects. This balance can create controversy between parties supporting a project and those opposing it over issues of ecological protection.

# 6

# Planning and Assessment

## INTRODUCTION

This chapter addresses the intersection between ecological systems and social systems by discussing the current practices of planning and assessing road projects, the environmental component in assessing and planning, and the possible solutions for integrating the two systems more effectively. The first section describes current assessments and planning processes and then describes planning at different spatial and temporal scales, different types of planning, and the interface between the planning process and the different social groups. The second section addresses issues and problems with the planning processes, and the third describes an array of ecological evaluation tools and methods used in planning and project development, including rapid assessment methods. The fourth section addresses some potential solutions, including a conceptual framework for integrating environmental factors along with other considerations in the planning phase.

## STATE OF ROAD PLANNING AND ASSESSMENT PROCESSES

Impacts from transportation systems and associated projects occur at several levels of environmental organization—from the individual organism to the landscape. Therefore, planning at the transportation system level to the individual project level requires a look at all components of the area being studied. However, as noted in Chapter 3, consideration of environmental issues is usually done at subregional scales. This perspective has resulted in the need for various levels of planning with increasingly finer levels of study to develop the most environmentally sensitive transportation system. The ways in which laws and regulations influence environmental assessment and policies regarding roads are similar to the ways they influence other land uses (Haeuber and Hobbs 2001). The lack of assessments at larger scales is due, in part, to a lack of an overview of multiple decisions and assessments at smaller spatial and temporal scales. This gap may be addressed in a cumulative impact assessment but is rarely done in practice. Hence, regional scale or larger assessments on ecological issues, such as species movements, large habitats, and aquifer, tend to be ignored by assessment at smaller scales.

## Key Elements of the Planning Process

### Coordination and Collaboration

Coordination and collaboration are required so that transportation agencies can provide the opportunity for input at all phases of planning and project development. Historically, resource agencies have had limited resources for input early in the planning process. This problem has been alleviated in some states where the transportation agencies are providing funding for resource agency personnel (see later discussion on Florida and Pennsylvania). Involvement of all parties early in the process helps to identify environmental factors in transportation plans and provides the groundwork for resolving issues early in the process.

### Stakeholders and Decision Making

Transportation is a critical component of our country's economy. Therefore, all segments of the population have a stake in transportation planning, and public-involvement programs of transportation agencies have expanded. The agencies are receiving input from diverse stakeholders representing many public and private interests. The environmental community has been active in transportation planning for several decades. The environmental resource agencies also are devoting greater resources to transportation planning and project development.

Stakeholder involvement in examining ecological impacts of roads has had some success. For example, incorporating stakeholder knowledge and understanding into decision making regarding road de-icing revealed tensions and inconsistencies between the mission and the operation of the institutional system as well as inadequate communication among various elements of the management system (Habron et al. 2004). However, the involvement of stakeholders facilitated understanding of the complex institutional system and helped to identify possible areas for improvements in development and implementation of watershed models. In other situations, local or even regional stakeholder groups are effective in helping to define long-term goals for an area where road development, use, or upgrades occur (e.g., MacLean et al. 1999).

## Federal Role in Planning

Under current laws and programs, there is a limited role for the federal government in planning that affects both transportation and ecological resources. Often such planning is conducted at the local and state levels rather than at the federal level. When federal agencies are involved, they may have little oversight or approval roles, such as approving Clean Air Act (CAA) conformity for a transportation plan or permitting Clean Water Act (CWA) impacts for particular transportation projects. Federal agencies may have expertise in particular ecological resources and may coordinate with state and local agencies on many levels. More frequently, however, there is little coordination of information and resources that would be helpful for ecological planning associated with transportation planning. Federal agencies are involved in natural-resource planning, however. The services are required to develop recovery plans for federally listed species, and other planning efforts are conducted for forests and other natural resources. Those services could provide opportunities for coordinated planning efforts.

## State and Regional Planning

State and regional transportation plans focus primarily on transportation needs, reflecting local choices for quality of life, mobility, growth, land use, and economics. Virtually all the rules and guidance for transportation plans include consideration of a variety of factors, including environmental factors, but the main emphasis is on meeting transportation needs. Traditionally, these plans involved efforts to keep the level of service of the transportation system compatible with growth. Although the planning process that results in state and regional plans includes consideration of environmental information, environmental issues generally are not a major part of determining the content of these plans.

For many reasons, the transportation planning stage rarely involves detailed consideration of siting of transportation projects, and there is a perception that most environmental issues can be addressed when a specific project location is viewed in more detail. There is also little information concerning ecological resources available for the transportation planning process. Relatively few locations have resource plans or consolidated information on ecological resources that could assist early-phase transportation planning. Some information on environmental resources to avoid—such as parks, refuges, and endangered species habitat—is considered during planning. Other information on ecological resources might not be available or might not be in a form that is useful for the planning scale.

To the extent feasible, transportation planning processes should use all available information on ecological resources to anticipate the potential impacts of the transportation plan on the ecological structure. That would require a number of shifts in emphasis and policy. In some states and regions, natural-resource plans address regional needs for certain resources, such as watersheds and certain migratory or endangered species. For example, some regions in California have multispecies habitat-conservation plans. In other instances, information is available in state, federal and other natural-resource management agencies, although perhaps not assembled in a plan. Obtaining useful information can involve considerable interagency coordination.

The transportation planning process will then also need to make commitments to address the ecological information at the planning stage. For example, if there is a natural-resource plan or other information describing a terrestrial migration route, it would be prudent for the transportation plan to consider whether a road planned within that migration route could be relocated. Where roads that cross streams must be improved, the transportation plan could consider and address the prospect of replacing culverts with bridges or replacing older bridges with wider spanning bridges, each of which would reduce the impacts of the road structure on the waterways. Taking up these kinds of considerations at the planning stage will have benefits of more accurately projecting expenses, as well as reducing controversy at the individual project stage.

## Long-Range Transportation Plans (Large Scale)

As introduced in Chapter 5, a long-range transportation plan (LRTP) must be developed for each state. This plan is reviewed annually and updated as necessary, based on air quality considerations to meet CAA standards (see Chapter 5). The development and review of this plan provide opportunities to examine environmental factors in relation to the actual transportation corridors as the plan moves forward. However, these plans are based on other factors (described in Chapter 5), such as the level of service for highways. Other than considering air quality, most LRTPs do not address environmental impacts in detail.

LRTPs exist at the state, local, or regional levels. Local and regional LRTPs are developed by metropolitan planning organizations (MPOs), consisting of county and local government representatives. In some cases, the state transportation plan or state LRTP combines various local or regional plans. For locations outside MPO jurisdiction (rural locations), the state is responsible for developing the LRTP. Other governmental entities may be involved as well, such as state legislatures, state transportation commissions, or other entities.

Although some states are trying to broaden the range of considerations in LRTPs, environmental factors, such as wildlife, and other natural-resource plans are usually not considered. Rather, LRTPs usually contain a "wish list" of desired transportation projects viewed as important to economic growth. LRTPs set priorities and must identify funding sources for all listed projects, thus keeping the "wish list" realistic.

## Transportation Improvement Plans (Medium Scale)

The transportation improvement plan (TIP) is the short-term, generally 2-year plan under a restricted budget that covers priority implementation projects and strategies from the LRTP. At the TIP stage, the state and regional or local entities determine which projects should be funded and proceed within the shorter time frame. Environmental concerns are not always comprehensively addressed in a TIP. Air quality issues, especially in nonattainment areas, must be addressed under federal law (see Chapter 5). This stage of planning provides the opportunity to revisit any environmental considerations identified in the LRTP to determine whether conditions have changed to make potential projects more or less environmentally compatible. TIPs also offer the opportunity to identify projects that environmental sensitivity dictates should be reevaluated before being included in the TIP work program.

## Project Planning (Fine Scale)

Projects identified in the TIP move on to project development, which includes environmental studies. This step is usually the first occasion when detailed (comprehensive) potential environmental impacts associated with a planned project receive environmental analysis; yet, some analysis should be done earlier. Project development studies have become thorough and consider all applicable state and federal laws. Even local laws, such as tree ordinances, are included in the studies.

Unfortunately, these projects have advanced to the point in the project-planning process that options to eliminate environmental impacts only include avoidance, minimization, and compensatory mitigation. For impacts with legal consequences, mitigation is included in the project studies so that projects can receive approval and permits. Ideally, modifications would be made earlier in the planning process.

## TYPES OF ENVIRONMENTAL PLANNING ACTIVITIES

Environmental planning related to transportation occurs at federal, state, and local levels. This section describes some of the more important planning activities as they relate to transportation. Many challenges that transportation agencies encounter result from a lack of coordination among the various planning activities that are often legally required. The challenges identified in the following discussion will probably persist until environmental factors are considered in all of these planning activities and the activities are coordinated. Small-scale environmental issues can be addressed accurately only at the project level—for example, how to span a stream. Determining whether a proposed road project can avoid streams or minimize the number of stream crossings might be more appropriate at a broader planning scale. Consideration of environmental issues at the planning stage could be facilitated if the various options for avoidance, minimization, and compensatory mitigation and rationale for the preferred option were better communicated to stakeholders. For example, at the planning stage, there might be several broad corridors considered for a road, and a preference might be expressed for the one that has fewest impacts on wetlands. Nonetheless, even within that broad corridor, wetland impacts will occur, but recognition of the avoidance steps taken in early planning may help to reduce objections during project review.

Systems planning takes a top-down and broad-scale perspective, whereas project planning deals with a specific project. Many governments and nongovernmental organizations (NGOs) have other planning activities that can be integrated with transportation planning. These activities include natural-resource planning, growth management, land-use planning, and economic development planning.

## Systems Planning

The state transportation system provides the opportunity to address the larger issues associated with environmental impacts. Systems-planning studies examine a multimodal transportation system and the future transportation needs at the state and interstate levels to arrive at a systems plan that is used to develop the LRTP. At this scale, environmentally sensitive corridors can be identified to determine whether moving forward with a project is in the best public interest. It is also possible to address the cumulative effects of projects on the landscape and to find means to avoid, minimize, or develop mitigation strategies for environmental impacts. This broad-scale perspective is especially important when evaluating impacts on wildlife connectivity for animals requiring large areas for maintenance of their populations.

If streamlining the environmental process is to be successful, environmental concerns associated with transportation plans and projects should be identified early so that project approval can move forward quickly. The goal of such proactive consideration of potential environmental issues would be to retain flexibility in avoidance and mitigation. Identification of issues and impacts at the systems-planning stage would allow the time and opportunity to resolve such factors before including a project into a plan.

Assessment of an environmental decision in a plan is a key part that is often overlooked (Bergquist and Bergquist 1999). Even though transportation plans (LRTP and TIP) are revisited annually, an important aspect of the planning process is that plans are frequently changed. Therefore, it is important that stakeholders continue to be involved at all phases of project planning, which can take many years and much effort.

## Project-Planning and Ecological-Impact Assessment

Project-level planning and ecological-impact assessment studies are done for those projects identified in the TIP when they are advanced to the design phase. This stage can be simple or complex. Projects that are judged not to have significant impacts can be categorically excluded from environmental assessment, and those that have potentially significant environmental impacts require more extensive studies. Level of documentation and necessity of associated studies are determined as project alternatives are developed and as issues or impacts are identified.

The National Environmental Policy Act (NEPA) requires study of social, economic, and environmental impacts of transportation projects that involve federal actions (such as financing, approval of connection to interstate system, and permits). This requirement has resulted in standardized procedures during the project development phase for documentation of potential environmental impacts of federal actions by the Federal Highway Administration (FHWA). The state transportation agencies and their consultants carry out these studies. The degree of study is based on potential social, economic, and environmental impacts. As described in Chapter 5, these studies lead to documentation in one of three forms: (1) categorical exclusion (CE), (2) an environmental assessment and finding of no significant impact (EA/FONSI), or (3) an environmental impact statement (EIS). Determination of the form of report is based on the significance of potential impacts and is arrived at jointly by the state transportation agency and FHWA. If there is a question of significance of impact or controversy associated with a project, an EA/FONSI or an EIS is developed. Over 90% of state transportation projects are processed as CEs, meaning they are deemed to have minor or no impact on the environment, as most of the projects involve expansion or modification of existing roads. Approximately 3-4% are processed as an EA/FONSI, meaning that the assessment resulted in a finding of no significant impact and the environmental issues were resolved. Less than 5% of transportation projects require an EIS meaning they may have significant ecological impacts or controversy (Forman et al. 2003). All projects require environmental study of potential impacts, but the degree of investigation is done on a case-by-case basis.

A recent study of highway practices found that, for the most part, states are successful in processing their environmental documents (EA and EIS) through a discovery process involving expert input and through coordination of the resource and regulatory agencies and the various public groups (Evink 2002). However, project delays can result because of environmental issues that require lengthy coordination.

The studies conducted during the project development phase are based on more specific information about potential project alignments and features than those conducted earlier. Therefore, the assessment of potential impacts is more complete during project development than at earlier planning stages. Problems can arise from the lack of information about associated environments. The need for long-term studies of potential impacts to the environment, such as those for a rare species, can cause unanticipated delays. Lengthy coordination with public-land managers for compatibility with land-management plans may be necessary. However, Evink (2002) found that most transportation agencies are trying to address environmental considerations. In that study, a survey indicated that Florida, Washington, Montana, and California, among others, have to finance the collection of basic environmental information on public lands before environmental analysis can be completed. Some agencies even finance the study of basic environmental factors, such as rare species occurrence, wetland functioning, and water quality, if unknown, to complete their environmental analyses. Often, these environmental issues might have been better resolved had they been identified in the early planning stages through a more complete environmental screening.

Although the NEPA document requirements are well defined, such clarity is not the case for assessment of environmental impacts associated with transportation projects. There is substantial latitude and flexibility in methods for the identification and mitigation of environmental impacts, and as a result, analytical quality varies among environmental documents developed by transportation agencies. The quality of the documents largely depends on the expertise of the scientists both within and outside the transportation agencies that are involved in the development of the documents. Therefore, a standardized analytical method could improve the overall quality of the assessments contained in the NEPA documents.

If issues and impacts are identified in the systems-planning process, the environmental impacts assessment and project development phase should provide sufficient details for resource agency signoff or the opportunity for early resolution of impact issues. FHWA recently published guidelines for better integration of transportation planning and the NEPA process (FHWA/FTA 2005). The project development and EIS phase are the least desirable stages to resolve complex environmental issues because project schedules often compress time frames.

## Related Planning Efforts

### Natural-Resource Planning

Natural-resource planning is conducted by federal, state, and local agencies and by NGOs to develop management plans for lands under consideration. Most of the agency plans are legally required. Coordination with the public and with other planning processes is also required for many of these plans. However, resource limitations and the complexity of the planning process often lead to little or no coordination among such plans. In some cases, resource management plans are not completed by resource agencies or are delayed because of limited resources. Hence, transportation agencies often find that they are trying to resolve resource and transportation planning issues without the needed input of plans from other agencies. This problem can lead to a long coordination process and project delays.

### Growth-Management and Land-Use Planning

Similar to natural-resource planning, local growth-management planning and land-use planning are ideally combined in a coordinated effort that considers future growth and land-use patterns at the local level. Consideration of environmental factors in local growth-management planning is an important step in the process. Historically, economic, social, and political factors have dictated local growth and land-use planning at the expense of environmental factors.

Infrastructure required by local growth-management and land-use planning is designed by transportation planning. The result is traffic patterns generated by anticipated growth. When environmental factors are not adequately addressed in the local plans, the transportation agencies have problems resolving these environmental problems during project development. These plans also suffer from inappropriate scale issues and lack of detail.

**Economic-Development Planning**

Economic-development plans interact with local growth-management and land-use planning. Transportation is a critical factor for many businesses. Therefore, implementation of a sound transportation system is important to the economic health of a state. Economic drivers receive high priorities in planning at all levels of government. When transportation facilities, such as highways, ports, and airports, are needed in environmentally sensitive areas to facilitate development, transportation agencies often have difficulty constructing such projects. A more desirable approach would be the resolution of environmental issues during regional planning. These planning efforts offer an important opportunity to help support improved environmental considerations in transportation planning.

## TRANSPORTATION ECOLOGICAL ASSESSMENTS

Because of the decentralized nature of transportation planning and project development, assessment of impacts (social, economic, and environmental) takes place at different levels of government. Planning has emphasized the idea that local planning would best address local transportation needs. Therefore, counties, municipalities, regional transportation authorities, and MPOs in concert with state transportation agencies have identified transportation needs and planned transportation improvements to address those needs. The motivations for these improvements have largely been social and economic.

The objectives of environmental assessment at the planning phase of transportation development are different from those at the project development phase. During the planning phase, possible projects are studied in relation to their ability to address transportation needs. At that point, specific project features are often not well defined. Concepts are thus studied in broad terms.

Most projects in transportation plans are improvements to existing facilities, and there is an existing alignment or structure with associated environmental features. A small percentage of the transportation planning in the United States involves a new facility at a new location.

One of the assessment objectives during the early planning stage, such as during the LRTP, can be the identification of environmentally sensitive features associated with proposed transportation improvements.

Recent guidelines published by FHWA (2005) suggest how such identification might be accomplished. Early identification of potential environmental impacts would eliminate especially problematic projects or at least allow for early coordination to work out environmental issues associated with proposed actions.

A number of agencies from the federal to local levels are developing the ability to evaluate environmental concerns early (see Boxes 6-1 and 6-2). Checklists of key questions can help resource managers identify the places and issues of greatest concern (e.g., Box 6-1). These questions relate potential environmental impacts of a proposed project to overarching goals, conservation interests, and regulations. The questions force the decision maker to think about the future for the region by considering the project in view of conservation plans and adjacent land uses. The questions in such a checklist can be answered more readily if data are accessible in a spatial data base (e.g., Box 6-2). These efforts are primarily spatially based macro-scale examinations of environmental features in the area of the prospective transportation improvements. However, there is no standard assessment method required for evaluation of environmental concerns in transportation planning or project developments, and necessary data are not always available at this stage.

## TOOLS AND METHODS FOR ASSESSMENT

Development of rapid assessment methods has been attempted for wetland functional analysis. The hydrogeomorphic method (HGM) of wetland functional analysis by the Corps of Engineers (Corps) is the latest effort (Brinson 1993). The HGM approach classifies a wetland based on its setting in the landscape, its source of water, and the dynamics of the water on-site. HGM has many useful applications in functional assessment, but most variables incorporated into the assessment models remain measures of wetland structure rather than processes, perhaps because of the cost of doing the functional assessment and the available staff (NRC 2001). Although methods for a few wetland types have been implemented, they have not been rapid assessment methods. Furthermore, the habitat evaluation procedure (HEP) developed by U.S. Fish and Wildlife Service (USFWS) in 1980 is not a rapid assessment method and has not become a standard assessment tool. The Pennsylvania Department of Transportation developed a community-based, landscape-level terrestrial mitigation decision-support system (Maurer 1999) that

**BOX 6-1** Example of a Checklist That Can Be Used in the Rapid Assessment Process: The Florida Environmental Technical Advisory Team

Rapid assessment of the potential for impacts of transportation projects will require greater interagency coordination in the early planning phases of transportation development. Florida is using a team approach (environmental technical advisory team [ETAT] made up of planning, consultation, and resource protection agencies) to conduct early screening of proposed transportation projects. Because more than 90% of the transportation projects being developed are processed as "categorical exclusions" (Forman et al. 2003), the team approach to early screening could resolve environmental issues in the majority of projects. Florida's screening process is a GIS-based system that uses habitat characterization and a series of ecologically related questions to determine the significance of the potential project impacts on the environment—in Florida's model, social, economic, and ecological considerations are considered in the process. Using only ecological concepts, the team considers the following types of factors early in the planning process to rapidly assess the potential for impacts and to ultimately decide whether the project can be moved to categorical exclusion or whether it requires further ecological study and coordination.

- Does the corridor meet state and federal goals and plans for the study area?
- Does the proposed corridor cross or adjoin lands acquired or planned for acquisition as public conservation lands?
- Does the proposed corridor cross or adjoin lands identified as important for wildlife in natural-resource management plans?
- Does the proposed corridor cross or adjoin federally defined critical habitat for a federally listed species?
- Does the proposed corridor cross or adjoin habitat other than critical habitat used by a state or federally listed threatened or endangered species?
- Does the proposed corridor cross or adjoin a 100-year floodplain of regional or state significance?
- Are proposed drainage alterations consistent with watershed management plans for the affected basins as defined by responsible water management agencies?
- Are there wetlands of state or regional significance in the corridor area?
- Does the proposed corridor cross or adjoin areas with natural resources of regional significance?

- Does the proposed corridor cross or adjoin a watershed that includes special designation areas, such as wild and scenic rivers and national marine sanctuary?
- Does the proposed corridor provide new access or increased access to a barrier island or involve an area identified as a Coastal Barrier Resources Act unit or is it located within a coastal high hazard area?
- Does the proposed corridor provide new access or increased capacity to a nonurbanized area?
- Does the proposed corridor induce or encourage land uses that are not contemplated by the adopted comprehensive plan?
- Does the corridor provide new or expanded access to environmentally sensitive areas?
- Will the proposed corridor support sound planned development?

In the Florida model, after consideration of these initial screening questions at the beginning of the LRTP process, the ETAT changes roles as the project progresses, providing further guidance and support should the project progress into the NEPA phase. For the approximately 10% of projects that cannot be categorically excluded, the screening process identifies the major environmental concerns that will need to be addressed during the process, thereby helping to accelerate conclusion of coordination and environmental documentation.

---

**BOX 6-2** Example of a GIS-Based Framework for Assessing Ecological Conditions in the Southeastern United States

The southeastern ecological framework (SEF) (developed by the University of Florida in 2001) is a geographic information system (GIS)-based analysis that can identify ecologically significant areas and connectivity in the Southeast. The states included in the project are Florida, Georgia, Alabama, Mississippi, South Carolina, North Carolina, Tennessee, and Kentucky. The approach is designed to meet the U.S. Environmental Protection Agency's (EPA's) goals of gathering and disseminating information pertinent to the ecological condition of a region. The framework focuses on implications for biodiversity protection, but the approach can be used for clean water and air provisions and global-change risk reduction. Data layers include conversation lands, hydrological features, wetlands, land

*(continued on next page)*

cover, potential habitat for species of interest, shellfish harvest areas, and the 100-year floodplain.

The development of a regional spatially explicit database and tools to access the information facilitates collaborative decision making, which is becoming a part of the process of environmental protection. Having spatially explicit data available for a region allows broad-scale implications to be considered in decision making. The spatial implications are important to consider because the growth of human populations and transportation systems, as well as economic development, are fragmenting natural landscapes. The need to consider connected actions and cumulative effects and to understand the geographic context where the actions will occur argues for an approach to project planning that considers a larger geographic area than that usually covered by a single project. The SEF provides a schema and contributes to the environmental decision-making capability of local communities and regions. The approach is useful for such diverse needs as community planning efforts and NEPA analysis. The SEF has been used to examine potential impacts of proposed roads.

A valuable aspect of the effort to create the SEF is the coordination across federal and private sources of natural-resource data. For EPA to continue to align its programmatic efforts with performance goals at regional and national landscape levels, it will need to rely on other state and federal agencies for sources of data, models, and expertise. Conservation organizations, such as the Association for Biodiversity Information and the Nature Conservancy, will also be a valuable source of data on locations and status of ecological systems, vulnerable species, and special sites of biodiversity significance. The use of the Southeast Natural Resource Leadership Group by the SEF work group is a model for coordination efforts. In addition, the effort to create a regional model to advance the management of the ecological framework is a novel and important step forward.

has some characteristics useful for rapid assessment. The Florida Department of Transportation has also developed a decision-making checklist based on geographic information system (GIS) data that could be used in a standardized assessment method (see Box 6-1). None of these efforts has led to a national standardized assessment method.

Tools for in situ monitoring, remotely sensed monitoring, data compilation, analysis, and modeling are constantly being improved. Advances in computer technology mean that tools for data collection and analysis are quickly getting into the hands of practitioners. However, because data are usually insufficient at the local level, remote-sensing techniques for data collection must be used when possible.

As the need for more strategic and broad-scale planning efforts to minimize the impacts of roads on ecological systems becomes apparent, various tools are being used to address the need. For example, quantifying the overall impact of new-road development on biodiversity and estimating the risk to biodiversity strongly depend on the availability, accuracy, reliability, and resolution of national data on the distributions of habitats, species, and development proposals (Treweek et al. 1998). Remote-sensing data are often useful to such analysis. Most estimates of land cover come from remotely sensed data. Data from Landsat satellites are particular useful for characterizing land cover, vegetation biophysical attributes, forest structure, and fragmentation (Cohen and Goward 2004). Satellite remote-sensing data can be used to develop maps of land cover at different times to assess changes over time in patterns and processes (e.g., Franklin et al. 2000). Monitoring such changes can be important in ecological monitoring programs at multiple spatial and temporal scales. In addition, broad-scale measures of species occurrence and behavior often rely on acoustical and video-monitoring tools or other type of remote measures of species (e.g., Sherwin et al. 2000).

Thresholds may be used to depict abrupt changes in ecological impacts, but in some cases, thresholds do not occur. For example, although excessive loading of fine sediments into rivers can result from road construction and such particles degrade salmonid spawning habitat, the linear relationship between deposited fine sediment and juvenile steelhead growth suggests that there is no threshold below which exacerbation of fine-sediment delivery is harmless (Suttle et al. 2004). Hence, any reduction in sediments could produce immediate benefits for salmonid restoration. However, thresholds can be important aspects of indicators related to road effects on habitat fragmentation (e.g., Gutzwiller and Barrow 2003) or habitat degradation (e.g., Johnson and Collinge 2004). The response of European three-toed woodpeckers (*Picoides tridactylus*) to different amounts of dead wood in a boreal and a sub-Alpine coniferous forest landscape in Switzerland was related to steep thresholds for the amount of dead standing trees and to a high–density road network (Butler et al. 2004).

## Issues, Gaps, and Problems with Assessments

Several issues arise in the current approaches to transportation assessments. They include the lack of standardized methods, indicators, and data. Other issues are the determination of the assessment scales and the focus of the approaches on specific ecosystem components (such as an endangered species) rather than on an integrated assessment. Other issues arise around who is involved in the assessment process and how government agencies interact with other groups.

The lack of a standardized method is the result of several factors. Federal transportation regulations and guidance and Council on Environmental Quality (CEQ) guidance outline in detail environmental factors that need to be considered in transportation assessments; however, those factors vary among regions, and priorities for those factors have not been established. Indicators of environmental quality or degradation have not been well defined (NRC 2000b).

Discussion continues on the temporal and spatial scale necessary to address potential impacts, especially secondary and cumulative impacts. For example, one such issue that can be assessed at multiple large scales involves the consequences of roads on fire dynamics in many parts of the country, especially in light of ENSO (El Niño/Southern Oscillation) and other broad-scale sources of climatic variability. Another largely unaddressed issue is the interaction of road networks and biota movement in context of climate change. The necessary environmental data do not exist for large areas of the country, and the data that exist lack reliability and compatibility because data vary among regions, making subsequent analysis difficult.

The existing process is a combination of coordination and an expert-opinion approach to identify the environmental factors within a project. Local and state officials are required to hold public meetings to discuss proposed plans and projects. They are also required to coordinate the plans with other state and federal agencies. A weakness in the existing system is that the information available during early planning meetings is often so inadequate and incomplete that constructive dialogue about environmental factors is put off until the project development phase, at which point it is often more difficult to reach consensus on environmentally sensitive projects. Project concepts and designs have not advanced to the point that potential environmental involvement can be identified. Often, information about the natural environments encountered during planning activities is lacking. Another weakness is the lack of public, environmental agency, and NGO input because of lack of resources. As a result, state transportation agencies have funded resource agency staff to obtain early input (Evink 2002).

Although the transportation field is reaching a point where a national rapid-assessment method for evaluating transportation-related impacts on the environment might be developed, there are still many obstacles. First, the basic biological information needed is lacking in many areas of the country. Next, there is no central source of such information when it is available. As a result, obtaining the information requires a major coordination effort by the transportation agency. EPA is working on developing such a centralized data set (Box 6-2; Durbrow et al. 2001). Often, the transportation agencies lack resources to generate missing information and must contract with other agencies or consultants to develop even the most basic biological information. An example is the basic biological information about an endangered or threatened species and their distribution. Lack of funding in the state and federal agencies responsible for wildlife management has led to a lack of management plans for habitat and wildlife in some areas of the country. Furthermore, additional work is needed in developing and testing models that address transportation impacts on ecological systems.

A rapid assessment method could be developed to evaluate the impacts of transportation actions on the environment and to generate mitigation strategies. It would interface GIS data with analytical models and empirical data to determine potential impacts under different scenarios. However, development of such analytical tools has not been an easy task, given the lack of data, mixed data quality, need for centralized data sources, and lack of resources for the development of data and techniques as well as scientific debate on measurement techniques. Rapid assessment components continue to be developed so that they can be implemented, but coordinating a rapid assessment method with the necessary participants may become overwhelming, as was the case when developing wetland models with the Corps-sponsored HGM. The original functional assessment procedure was modified to meet different needs, and most deviate from the original intent, premise, and design (Brinson 1995, 1996). If a standardized rapid assessment method becomes a national imperative, the following are some of the actions that will be necessary to produce a universally acceptable product.

A factor limiting the development of a rapid assessment method has been the lack of uniformity of data collected at one scale of resolution, using standard means for measuring, sampling, and labeling. The need for uniformity is especially true of data useful in GIS. Data uniformity is lacking even among agencies in many state governments, let alone in the nation. (For example, soil data are inconsistent from state to state.) State and federal agencies and industry groups have been discussing for years what format and systems should be used, and some progress has been made. The lack of uniformity also applies to the quality of data available. Many different entities are developing data, and the quality of the data is dependent on the resources and expertise brought to the task.

Scientific debate continues about appropriate indicators of ecological attributes. EPA developed a framework that attempts to define these multiple attributes (Figure 6-1), Dale and Beyeler (2001) developed criteria for indices, and the NRC (2000b) proposed a set of national ecological indicators and described the scientific rationale for them. There are six essential ecological attributes: (1) Landscape conditions are the extent, composition, and patterns of habitats in a landscape context. An example is the status and change in extent of a forest system. (2) Biotic condition is the viability of communities, populations, and individual biota. Examples include the presence and population trends of rare and common species. (3) Ecological processes are the metabolic functioning of ecological systems, such as energy flow, element cycling, and the production consumption and decomposition of organic matter. (4) Chemical and physical characteristics are the physical conditions (for example, temperature) and concentrations of chemical substances (for example, phosphorus) present in the environment. (5) Hydrology and geomorphology are the interplay of water and land form in the environment. Soil erosion and changes in streamflow rates fall into this category. (6) Natural disturbance regimes are the pattern of discrete and recurrent disturbances that have shaped an ecological system. Examples include wind throws, floods, and fires. Often, the relevant indices depend on the project questions and options (Figure 6-2).

Finally, the lack of mandatory legal standards or requirements for many ecological features, other than air quality, water quality, and wetlands, complicates the issues of whether and how a standard assessment method should calculate impacts or set metrics to assist with mitigation choices.

On a more positive note, scientific and technological development is leading the nation in the direction of a better understanding of the environment and the impacts of human activities. This advancement is largely the result of computer technology and the associated capabilities for data storage and analysis. Models using GIS data hold great promise and are being developed to look at diverse sets of parameters to support

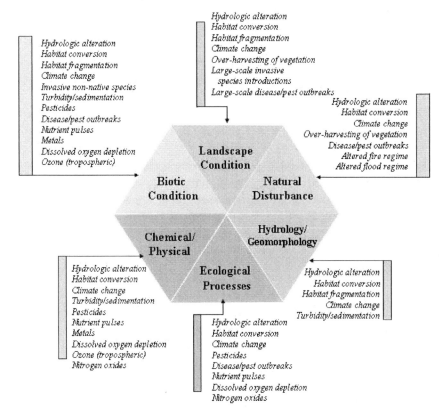

**FIGURE 6-1** EPA ecological framework: essential ecological attributes. Source: EPA SAB 2002.

## SAB Proposed Terminology

| Administrative Measures | Stressor Measures | Exposure Measures | Effects/Condition Measures |
|---|---|---|---|
| e.g., number of TMDL's completed for metals | e.g., concentration of metals in a stream | e.g., concentration of metals in stream biota | e.g., composition and diversity of stream biotic community |

**Least directly related to Environmental Outcomes**        **Most directly related to Environmental Outcomes**

**FIGURE 6-2** The spectrum of environmental performance measures. Abbreviations: SAB, Science Advisory Board of EPA; TMDL, total maximum daily load. Source: EPA SAB 2002.

decision making (see Dale 2003). An example is the landscape ecological analysis and the rules for the configuration of habitat (LARCH) model, an expert systems model that uses thematic mapping to examine ecosystems and network populations to compare different landscape scenarios to support decision making in considering project alternatives in The Netherlands (Bank et al. 2002). For mitigation, models are being developed that identify wildlife habitat linkages for planning wildlife crossing structures (Clevenger et al. 2002a).

Modeling tools are being developed to meet the needs of environmental assessment. The use of models is based on the concept that models can synthesize the current understanding of the environmental repercussions of an action. Models offer the opportunity to deal with complexities and interactions without being overwhelmed by details. In particular, future modeling should deal with scales, feedback, and multiple criteria for environmental protection (Dale 2003). Models can integrate processes that operate on various temporal and spatial scales. Incorporating feedback from different aspects of the environment that operate at different scales is one of the biggest challenges of interdisciplinary research. Environmental protection has been constrained by efforts to meet one single criterion (for example, protection of a single species or keeping water quality above a given standard). Developing approaches allow simultaneous consideration of several criteria and their interactions. The use of ecological models by decision makers also requires some improvements. These include establishing closer communication between environmental scientists and modelers, developing clearer definitions of the land and resource management issues, using models to enhance understanding, and exploring alternative future conditions (Dale 2003).

Some transportation agencies have begun to use these techniques in transportation planning and project development. Natural Heritage Programs and other state agencies are providing GIS data. A few states' transportation agencies have even contributed to the development of these data layers. Florida has completed GIS layers for many features of the state and is using the information in systems planning and project development studies (Carr et al. 1998, Kautz et al. 1999, Smith 1999). Florida is also developing models using the GIS data to identify potential impacts on environments, given different types of factors (Box 6-1). These data are then used in coordination with resource agencies and the public to arrive at the transportation plans and subsequent projects. An interagency effort (the Southeastern Ecological Framework) has been formed to develop this type of data for the entire southeastern United States (Box 6-2).

The state efforts are in response to the need for more complete disclosure of information related to transportation planning and project development to streamline the approval process. The internet is contributing substantially to sharing this information. State transportation agencies are posting planning and project development information on the web as part of their public involvement programs, providing a convenient opportunity for agencies, organizations, and the public to have input on a project. Another promising area from the internet is international sharing of information. Scientific information on the impacts of transportation on the natural environment is being gathered and posted at sites by several of this nation's transportation centers (for example, the Center for Transportation and the Environment and the Western Transportation Center) and internationally by the multinational research committee COST[1] effort in Europe, along with a host of university, governmental, and private sites (Evink 2002, Bank et al. 2002).

To develop standard methods, a number of steps need to be taken. First, resource agency budgets would need a major increase allocated to the development of basic biological information in a uniform format. This step would have to take place at state and federal levels to develop the data and planning necessary for input to the system. This information could be made available through a web-based system. Second, a national effort to develop environmental GIS data at the local and state level in a uniform format would be necessary. Data layers that are useful to include are land cover, roads, utility lines, soil types, hydrological units, streams and lakes, public ownership lands, and patterns of past disturbances Although some of this work is already taking place, it cannot be used in all locations of the country. A web-based system of the data would be necessary for efficient sharing of the data.

Next, interdisciplinary groups of scientists, including personnel from resource and regulatory agencies, would be tasked to identify the

---

[1] COST is a European framework for the coordination of nationally funded research. A "COST Action" is a European Concerted Research Action based on a Memorandum of Understanding (MoU) signed by the Governments of the COST Countries wishing to participate in the Action. Each COST Action is identified by a number (e.g., 341) and a title (e.g., Habitat Fragmentation due to Transportation Infrastructure). Reference: http://www.cordis.lu/cost-transport/home.html.

analytical models needed for the various environments encountered by transportation projects—wetlands, aquatic systems, uplands, and so forth. Research would be needed to develop the analytical models. Models that include mitigation components would need to allow examination of costs and benefits of various alternatives. The approach would involve interdisciplinary groups who would develop an interactive analytical program that would form the core of the rapid assessment methodology. Approval by a wide audience would be necessary to make the effort fruitful in the end.

Some state transportation agencies, such as Florida's, have already proceeded along this path. However, progress is slow because of sparse funding, and lack of reliable basic data, completed land-management planning, and other information needed to make the system work. Getting consensus from resource and regulatory agencies has often been difficult because of the diversity of interests involved. Any increase in activity at the state and federal level would help provide the science to make a standardized assessment method possible.

## POTENTIAL SOLUTIONS AND IMPROVEMENTS FOR ASSESSMENT AND PLANNING

### Conceptual Considerations

Environmental assessments and planning activities involve some estimation of future events. These processes involve trying to predict environmental effects associated with a certain action, such as building or widening a road. Although broad, general outcomes are known, the ability to predict ecological outcomes is difficult and limited for many reasons. Ecological systems are complex; the scales of ecological structures and processes range widely; there are many competing ideas about the structure and functioning of ecological systems; and the lack of information and financial support to test ideas about ecological systems prevent resolution on a meaningful scale. Those problems can be applied to the social dimensions of assessment and planning. Social-dimension problems include the shifting and multiple values that society places on different aspects of ecological systems.

Another way to think about assessment and planning is that both are processes that address the consequences that may result from a proposed action. Several approaches have been developed to address and reduce uncertainties. Those include qualitative assessments by knowledgeable persons, narratives or scenarios, strategic gaming, statistical analyses of empirical or historical data, and causal modeling (NRC 1999). Expert opinions can be used in legal situations or when causal understanding or data are lacking. The development of narratives or scenarios has been used successfully to highlight emerging issues (for example, Carson 1962) or to roughly define feasible alternative futures that contain great uncertainty. Scenario-based planning has been used in business (Shell Corporation [van der Heijden 1996]), climate-change research (Raskin et al. 1998), and exploration of sustainable futures (NRC 1999). Empirical approaches that involve statistical analyses of existing data sets are dependent on extant information in forms amenable to analysis. On the basis of the amount and type of uncertainty associated with ecological issues, all of those approaches can be used to assess resource issues in transportation. A conceptual framework is proposed in the next section as a way to improve integration of ecological considerations into transportation planning.

## Conceptual Framework for a Rapid Assessment Method

The committee's proposed conceptual framework for a rapid assessment method systematically incorporates ecological factors into transportation decision making. Its rapid assessment approach is based on those that use basic ecological principles (Stohlgren et al. 1997, Stork et al. 1997, Dale et al. 2000, Dale and Haeuber 2001). Many aspects of this approach were field tested by the Pennsylvania Department Transportation (see Maurer 1999). In addition, the Florida Department of Transportation developed a checklist (Box 6-1) using GIS data that could be the basis for a standardized assessment. In this section, the committee takes the elements of existing methods, develops a framework, and proposes that a national method be adopted.

The goal of the rapid assessment method is to provide environmental information to decision makers at several stages in the process of making decisions so that the ecological impacts of roads can be evaluated. The ecological impacts must be considered in early decisions. However, such early inclusion is possible only if the ecological information can be provided in a timely manner. Streamlining of ecological information early into the decision-making process can reduce costs.

The rapid assessment approach offers new opportunities to better protect, restore, and enhance ecological systems affected by the construction, maintenance, and operation of transportation systems. At present, this method cannot be rapid, but it should become so as the approach becomes more familiar, as the required data become more available, and as the tools for data manipulation and analyses become more widely accessible. The committee's proposed framework calls for

• Consideration of ecological factors early in transportation systems planning and throughout project development, maintenance, and operations.

• Establishment of ecological goals and performance indicators along with transportation goals and performance indicators to meet ecological resource protection, restoration, and enhancement needs as well as transportation needs.

• Identification of the ecological context of the study area in the form of the mapping and characterization of specific features of the ecological system before the development of transportation systems plans, programs, and projects.

• Use of a context-sensitive solutions planning and design approach to explore a range of transportation solutions in lieu of a design and impact assessment approach.

• Use of ecosystem performance assessment models to evaluate the ecological performance of alternative solutions.

• Provision of information and involvement of the public throughout the process.

• Monitoring and evaluation of ecological protection, restoration, and enhancement during construction, maintenance, and operations.

• Ecological research to develop new baseline information, analysis methods, and tools.

• Use of ecological databases and models for managing baseline, monitoring, and evaluation information.

• Formation of multidisciplinary and multiorganizational teams that collaboratively conduct ecological analyses related to transportation issues.

• Expansion of the consideration of ecological factors beyond transportation project development for which ecological impacts have been addressed historically through compliance with NEPA and other environmental laws, regulations, and policies.

The committee's proposed conceptual approach is described as a five-step process (Figure 6-3) in the following paragraphs.

### Step 1—Establish a multidisciplinary and multiorganizational team

Before this process is begun, however, an analysis is needed of the probable effects of the project on the environment, based on considerations such as those discussed in Chapter 3. Thus, if the project is the construction of a new major highway, assessments would be conducted on the potential for habitat alteration; disturbances due to noise and light;

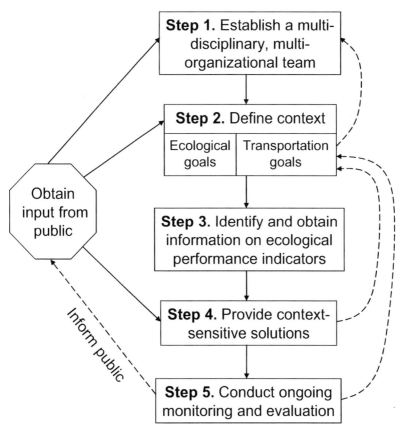

**FIGURE 6-3** Diagram of committee's proposed framework for a rapid assessment method.

fragmentation of habitat; interference with the dispersal of most organisms; facilitation of the spread of some organisms, possibly including exotics; a risk of deaths to mobile animals caused by collisions with traffic; possible changes to hydrological and sedimentation patterns; and the probable accumulation of contaminating chemicals, including de-icing materials in most regions of the country.  If the project is a bridge replacement, resurfacing or widening of an existing road, or similar activity, then a different and smaller suite of potential environmental effects will need to be assessed.  In addition to the road, the structures that will need to be identified include culverts, bridges, overpasses, on- and off-ramps, and causeways.  Once the preliminary but essential assessment has been done, the remaining five steps should be followed as described below.

The team should consist of planners, engineers, and environmental professionals who have experience with the particular conditions related to ecological impacts of roads.  The team members will develop and apply the method.  Depending on the makeup of the team, the method could be used at the project level or the system level.  It is critical, however, that the ecological impacts of roads be considered at several stages in the road planning process, especially at the early stages.  As the project evolves, team membership may change as different expertise is needed.  For example, if a road crosses a karst area (terrain usually formed on carbonate rock, characterized by deep gullies, caves, sinkholes, and underground drainage), a specialist may be needed.

## Step 2—Define context

Defining the context requires identifying ecological and transportation goals, ecological concerns, and spatial and temporal scales of interest.  The ecological goals should include protection of species of concern and habitats, specific ecological or geological features, and ecosystem services as well as transportation goals (for example, to develop a new road to meet transportation needs).  The Committee of Scientists (1999) argued that meaningful planning sets both long-term goals and desired future conditions.  Thus, the link between developing assessments and building decisions is in defining the desired future condition that includes both ecological and transportation goals.  Defining a desired future condition requires extended public dialogue, as it is a social choice affecting current and future generations.  As a future-oriented choice, a desired future condition seeks to protect a broad range of choices for future generations, avoid irretrievable losses, and guide current management and conservation strategies and actions. Nonetheless, given the dynamic nature of ecological and social systems, a desired future condition is also dynamic over time and thus is always revisited in the decision-making process during monitoring, external review, and evaluation of performance.

The public can be involved in all stages of setting goals. From a social perspective, desired future conditions are those that sustain the capacity for future generations to maintain patterns of life and adapt to changing societal and ecological conditions (Committee of Scientists 1999). Different parts of society and different stakeholders offer interpretations of both past and possible future conditions, reinforcing the importance of the deliberative process of collaborative planning.

Ecological assessments offer independent information about ecological conditions, against which differing perspectives can be compared. This process will provide a mosaic of explanations and perspectives, all of which need to be a part of the deliberation of desired future conditions and the implications for future choices of different ecological conditions (Committee of Scientists 1999). Although choosing a particular desired future condition is a social choice, this choice is still bound by biophysical limits. Thus, the desired future condition represents common goals and aspirations.

Defining ecological concerns also requires addressing legal and social issues. Laws that protect the environment are addressed in Chapter 5 and include NEPA, the Endangered Species Act, the CWA, the CAA, and many other national and local laws and ordinances. Social issues are typically context specific and may include the need to preserve certain places that are highly valued by the community. Thus, context definition draws on the experiences of all of the stakeholders.

## Step 3—Identify and obtain information of ecological performance indicators

Performance indicators for the desired future conditions are metrics that quantify the features for the desired outcome. The essential ecological attributes (see Figure 6-1) can be considered a checklist to help to develop these indicators. Some of these values may be available from existing data sources (for example, the southeastern U.S. framework; see Box 6-2). Characterization of the ecological system can be done by quantification of the selected indicators. However, the indicators can vary over time and space (Figure 6-4). When possible and reasonable, mapping of the performance metrics provides spatial attribution of the indicators.

Other experience with environmental assessments (Holling 1978; Walters 1986, 1997) suggests that dynamic computer models for re-source-managed applications can be powerful tools at this step. These models link data on chosen indicators with information on critical eco-system factors that influence these indicators. These models also include linkages to determine how alternative design or policy options would affect indicators. GIS may provide useful information on spatial rela-tionships and slowly changing ecological attributes but are not as effec-tive in understanding dynamic linkages. Another attribute of this model-ing approach is that it enables the technical team to assess feasibility of policies, which is critical to determination of context-sensitive solutions.

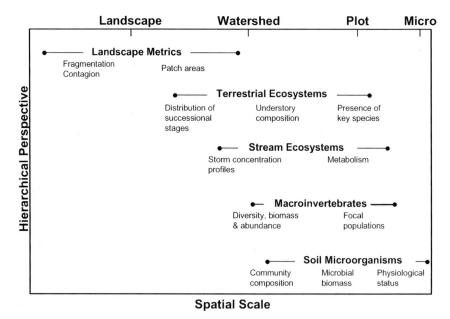

**FIGURE 6-4** A suite of indicators across scales is being adopted by the U.S. Department of Defense's Strategic Environmental Management Project. Source: Dale et al. 2002. Reprinted with permission; copyright 2002, Elsevier.

## Step 4—Context-sensitive solutions

Information generated by an ecological assessment can contribute to the understanding of ecological processes, but it cannot determine what choice is right. Informed expert opinion and public dialogue is essential to determine decision options. Hence, a first step for decision making is to develop the public forum for defining the desired future condition. The decision makers can use the options generated and projections of the ecological impacts of their implementation to plan and design context-sensitive transportation solutions.

Options for meeting the transportation goals can be considered in view of their potential effects on ecological conditions. Enhancement and mitigation can be built into the array of proposed solutions. A plan can be designed to convey the information of changes in performance indicators to planners, engineers, and decision makers. Modeling and assessment tools can be used to determine whether any solutions meet (or exceed) the ecological and transportation goals. If goals are met for a proposal, then that solution may be selected. If not, then the team can use the assessment information to determine whether the impacts from a solution are acceptable. If the ecological ramifications are not acceptable, then the transportation goals may need to be reevaluated, or if the transportation consequences are too severe, then it may be necessary to reset acceptable ecological goals. This proposed process may not reduce or eliminate all issues.

Regarding proactive actions to address ecological concerns, the effect of actions driven by laws and ordinances and social concerns can be very different from those designed merely to meet the short-term intent of the law. In addition, proactive measures can serve to speed development by considering all possible negative implications of a decision and implementing solutions to a potential problem early in the process (often using actions that are relatively inexpensive to implement at the early phase, for example, small relocations).

## Step 5—Ongoing monitoring and evaluation

After information from the rapid assessment process has informed decision making and a decision has been implemented, it is critical to continue monitoring and evaluating to determine whether the ecological and transportation goals are being addressed and to provide the opportunity for mid-course corrections (Bergquist and Bergquist 1999). When corrections are necessary, it may be useful to go back to step 2 of this process (Figure 6-3) and redefine the context, obtain new information on performance indicators, reevaluate context-sensitive solutions, and reinstate a new monitoring and analysis protocol. Unfortunately, existing laws and regulations do not require reevaluations, but such reconsideration often makes sense in terms of management and budgetary needs. Planning for ongoing monitoring and evaluation requires well-maintained and accessible databases and the ability to use performance indicators to track the ecological implications of natural changes and management actions. Data and indicators are critical components of this proposed framework. In addition, such reevaluation allows for corrections to be made to correct for deficiencies in performance and for ongoing communications with the public.

## Data and Indicators for Assessment

To conduct ecological impact assessments as part of transportation decision making, the transportation and ecological context of the area of potential impacts must be well defined and understood. These contexts are defined through the collection and analysis of baseline transportation and ecological data. The collection of such data usually involves a combination of acquiring existing data files and conducting supplemental field studies. The scale, quality, quantity, accessibility, and age of the existing data files on baseline conditions heavily influence the amount of supplemental field studies necessary to define the transportation and ecological contexts adequately.

Local, state, and federal transportation agencies over the past 100 years have assembled a tremendous amount of macro-scale and micro-scale baseline data about the transportation system. In the past 25 years, a concerted effort has been made to create national electronic transportation databases and database management standards to improve the quality and accessibility of the transportation data. With the advent of the internet and GIS, transportation agencies have developed web-accessible databases and decision-support systems that make high-quality transportation system data readily accessible for decision making at the local, state, and national levels. The FHWA in cooperation with the local, state, federal, and tribal transportation agencies through the American Association of State Highway and Transportation Officials have ongoing transportation data collection and management enhancement programs to continually update and improve the quality, quantity, and accessibility of data.

Although local, state, and federal regulatory and environmental agencies and interest groups have assembled and managed a tremendous amount of baseline ecological data over periods similar to the development of transportation data, the data are typically low quality, at a macroscale, and rarely adequate for ecological impact assessments. Very few transportation ecological impact assessments are completed with previously existing ecological data. Most often, the transportation agency environmental professionals must collect ecological field data at a microscale for their ecological impact assessments.

The timing and extent of supplemental field studies often drive the transportation project delivery schedule. In the case of ecological field studies, their timing and extent is usually heavily influenced by changing environmental and biological conditions that could result in several months to years of field study. In contrast, transportation field studies involve measurements that are not as heavily influenced by changing environmental and biological conditions, allowing them to be completed in weeks or months. The short time frame of transportation field studies may be inadequate to gather sufficient data on important ecological processes (such as animal migration) that may require at least 2 years to document adequately.

Ideally, high-quality ecological databases with macro-scale and micro-scale data would be readily accessible to transportation agencies to define the ecological context for ecological impact assessments. They would be developed and operated in accordance with national database standards. Such databases would probably substantially reduce the number of ecological field studies necessary to supplement the ecological databases. By reducing the number of ecological field studies, ecological impact assessment and transportation program delivery schedules could be shortened. The real world of databases, models, and GIS is quickly approaching this ideal goal. Already, emergency services use spatial databases to plan and implement civic actions. As these tools are adopted by municipalities and states, they can be used for planning and assessing ecological impacts of roads. For example, standard GIS data on road networks (for example, TIGER/Line data) could be interfaced with data models (for example, UNETRANS).

The typical ecological database information that would be used in defining the ecological context of the area of potential impacts includes

- Land uses, including consumptive and nonconsumptive uses
- Topography
- Soils
- Surface waters
- Groundwater
- Vegetative cover
- Plant species population locations, including threatened and endangered plants
- Animal species population locations, including threatened and endangered animals
- Important wildlife-movement corridors intersecting transportation corridors
- Important plant and animal species habitats, including threatened and endangered plants and animals
- Degraded plant and animal species habitats with restoration potential

Models need to be in place to manipulate these data in ways that are useful to address potential ecological impacts.

A hierarchical approach to assessing ecological condition was developed and tested by the Center for International Forestry Research (CIFOR 2004). This approach begins at the landscape scope and uses the broad-scale perspective to focus the assessment. Such an approach would be very useful to evaluate the ecological impacts of roads.

## SUMMARY

This chapter provides a review of current practices in transportation planning and assessment. Large-scale planning processes (such as LRTP) are only required to address air quality issues (such as attainment of standards) and generally do not address other environmental issues. Some states are including environmental information earlier in the planning process, thereby improving the coordination of environmental issues and simplifying the project development process. The science needed to improve early environmental assessment in the planning process is evolving. Models and methods are discussed with examples of some of the more successful approaches. Conceptual analytical approaches are presented and a framework developed. Weaknesses and information gaps needed to bring these concepts to reality are identified. To streamline environmental assessments, three steps must be taken: (1) A checklist addressing potential impacts should be adapted that can be used for rapid assessment. Such a checklist would focus attention on places and issues of greatest concern. (2) A national effort must be made to develop standards for data collection and then to collect appropriate environmental data (both spatial and temporal) across a wide range of scales. These data should be made accessible and available via the internet. (3) A set of models (such as GIS) must be expanded to allow those data to be used in assessments to address scale, feedback, and mixed criteria for environmental protection. Transportation agencies can continue to expand their roles as information brokers and to foster planning forums that integrate environmental planning and assessment across agencies, NGOs, stakeholders, and the public.

# 7

# Integrating Obstacles and Opportunities

## INTRODUCTION

The previous six chapters underscore the need for integrating environmental concerns into all phases of road development and the special challenges to that integration. In this chapter, some of the barriers to integration and some ways to eliminate or lessen those barriers are discussed. As suggested in Chapter 1, integration is critical in two areas: integrating across scales of space and time and integrating ecological and social concerns. The first section in this chapter briefly summarizes how issues of scale influence the understanding and management of road ecology. The second section reviews how environmental issues can be integrated with other social concerns. The chapter concludes with a section on facilitating integration in practice and suggests some alternative approaches and considerations for developing integrative solutions.

## ISSUES OF INTEGRATION ACROSS SCALES

Ecological systems self-organize across a wide range of spatial and temporal scales. They operate over spatial scales of millimeters to thousands of kilometers and temporal scales of milliseconds to millennia (see Chapter 1, Figure 1-9). The span of scales over which ecosystems operate is much larger than the span of scales evaluated in this report. In the evaluation of the ecological effects of roads and in the assessment and management options associated with those effects, the committee focused on spatial scales from meters to hundreds of kilometers. Temporal scales included in this report range from days to a few decades.

In confronting the broad range of scales covered by ecosystems and road systems, the approach has been to divide the scales into research, assessment, and management categories. As shown in Chapter 3, most research on, and understanding of, ecological effects covers small scales—road segments and corridors. The committee has inferred effects at larger scales (disruption of landscapes and spread of exotic organisms), but few studies cover these scales. In all of these studies, bounds are established that define a discrete scale range. Studies are bounded in space (for example, meter square plots to hundreds of hectares) and time (for example, days to years), whether they are field investigations or modeling-based inquiries. As with the understanding of ecosystems, the scales for road planning, assessment, and operation are also divided; long-range transportation plans (LRTPs) cover regions and decades, and state transportation improvement plans cover road segments and years (see Chapters 3 and 6). This division of scales has its roots in theory, as described in Chapter 3; Chapter 3 also discusses cross-scale effects and their relevance to road ecology. The remainder of this chapter focuses on integrating social (including legal, institutional, and economic) and ecological issues.

## Scales of Law, Planning, Assessment, and Financing

Federal laws apply to the lands within the United States, unless specified otherwise, and thus cover a wide range of spatial scales, from a very small scale (molecular level for water or air quality) to a national scale. State laws are restricted to the geographical extent of that specific state and may address environmental concerns in relation to roads. Even though these laws cover a conceptually wide area (the nation), the spatial scope is narrowed greatly during the planning and assessment process. For example, consideration of wetland resources is only applied to specific areas where wetland ecosystems occur. Similar restrictions are applied to endangered species consideration, generally applied to a specific population and habitat, regardless of wider distributions. Hence, there appears to be a consideration of the relationship of immediate (spatial scale) impacts on wetlands and on endangered species, but insufficient consideration of how those immediate impacts relate to impacts on a greater scale. This narrowing in scope of federal statutes requiring envi-

ronmental considerations occurs when road projects cross state borders—
that is, ecosystems cross state borders, but planning of road projects does
not.

One of the gaps identified by the committee is the lack of protection
of ecosystem goods and services at scales larger than the road segment.
A broader and consistent set of ecological concerns should be addressed
to provide that protection.

The current system of planning is conducted at two scales: the project level (e.g., transportation improvement program) and the regional or
state (LRTP) level as detailed in Chapter 6. Project level planning addresses certain environmental issues that are required by law or policy
and addresses them on restricted scales (up to watershed scales in some
cases). The understanding of ecological effects is restricted in a similar
manner to scales from small to intermediate (see Chapter 3). Almost no
studies of ecological effects at the national scale were found in the literature, nor were national-scale data sets. Regional and state planning addresses such issues as economics and population growth but rarely addresses ecological concerns. Hence, most of the ongoing road assessment and planning ignores ecological structures and processes that occur
at broad scales. Transportation-related issues at the national scale are
addressed by many groups, such as the Federal Highway Administration,
the American Association of State Highway and Transportation Officials, the surface transportation policy project, and nongovernmental organizations (NGOs), with few, if any, intersections with national-scale
ecological issues (for example, global change, biodiversity loss, and exotic pest biota). At smaller scales (regions to municipalities), more ecological knowledge becomes integrated. This situation suggests that a
better integration of transportation and ecology is needed across a wide
range of scales.

Deciding on an appropriate scale for assessment and management
of road effects on ecological systems is problematic. Bounds are established to analyze road effects. No matter where those bounds are set,
structures and processes at larger and smaller scales must be considered.
There are few opportunities in the current system (from planning through
project implementation) where these two scales are addressed. The key
question for each assessment and monitoring effort is determining the
spatial and temporal scale of interest.

Assessment and planning should be redesigned to fill these gaps in
scale. A multiscale approach will be required. Models and data sets

should be developed at different scales. Ecological effects occur over wide scales and cannot be addressed solely by scaling up or down on the basis of existing information or models. Experience suggests that multiple models should be developed to deal with the cross-scale dynamics of these systems (Holling 1978, Walters 1986).

## INTEGRATING ECOLOGICAL CONCERNS AND SOCIAL OBJECTIVES

The issues that arise in road planning, development, construction, and use are complex. One source of complexity is from the many competing ideas and agendas that exist among various people, groups, and agencies involved in these processes. These groups include governmental agencies at state, local, or federal levels, private consultants and experts, NGOs, academics, interested individuals, and the public. Recognition of the sources and nature of these perspectives can help in developing strategies to confront complexity while meeting goals of integration and streamlining. A set of processes needs to be developed to manage these complexities.

The technical contributors from the list above (those concerned with understanding and evaluating environmental effects of roads and applying that understanding) are all trained in an academic discipline. A source of integration complexities arises from differences among many disciplines that underlie their practices. For example, conservationists' ideas are rooted in understanding of ecology or biology; road engineers in mathematics and physics; planners in geography or architecture. The differences in paradigms, theories, methods, and practices among disciplines provide large gaps and problems in methods, approaches, and language. For example, ecologists or engineers can be very good at understanding and evaluating environmental effects but could be better trained at communicating to a wider audience or trained in alternative models of the realities of human behavior, organizational structures, and institutional arrangements. Another such example is when engineers or managers assume that the uncertainty of nature can be replaced by human attempts to control and stabilize ecological systems. It is not that these approaches are wrong but that they are partial and require more disciplinary integration (Gunderson and Holling 2002) and perhaps more broadened, cross-disciplinary training.

Another type of complexity occurs when a diverse set of social values held by a myriad of groups intersect. Road planning and design engage governmental agencies, conservationists, and other stakeholder groups. Federal agencies—such as the U.S. Fish and Wildlife Service, which has responsibilities under the Endangered Species Act and the U.S. Environmental Protection Agency (EPA) and the U.S. Army Corps of Engineers (Corps), which are charged with protecting wetlands—have specific mandates that are not overlapping and can come into conflict with other governmental objectives. Many states have similar agencies that are also engaged. NGOs and even sovereign entities (Native American tribes) are becoming more involved in lobbying, planning, design, and management of roads (for example, the Nature Conservancy and the California Department of Transportation). Current efforts of assessment and planning, reviewed in Chapter 6, describe how various groups are engaged. Yet difficulties persist and protract the planning and design process because of the number and types of agencies and stakeholders involved.

In addition to the number and types of competing interests, another obstacle to integration is the ways in which the groups interact and resolve differences so that actions can be taken. Differences can be resolved in a variety of ways, from discussions at workshops or public meetings to formal dispute resolution or lawsuits.

The set of laws and bureaucratic structures developed to implement policy appears to generate partial solutions to a myriad of problems faced in integrating environmental concerns into all elements of road design, construction, or management. Moreover, legal and bureaucratic frameworks are set up in a way that defaults first to administrative processes and second to legal institutions to resolve conflicts. In these approaches, integration of environmental concerns with other issues can be, but is not always, achieved. Often, this deficiency occurs because of a lack of understanding about the natural systems or because of unrealistic expectations about how natural systems behave. Sometimes, partial solutions generate new ecological problems. For some controversial projects, there is a clear understanding about the natural systems that will be affected by a transportation project, but existing institutions do not provide a way to address the concern, so project opponents must select from a small number of mandatory requirements, such as clean air conformity, for their concerns to be addressed. Hence, what appears to be needed is a new type of institutional arrangement or structure that is based on an understanding of the integrated nature of the social and ecological components of road ecology (such as step 1 in the committee's proposed assessment framework [see Figure 6-3]).

## TOWARD INTEGRATIVE SOLUTIONS

In this section, the committee presents some proposals for better integration of ecological concerns with all phases of road activity from planning through management. Previous authors (Forman et al. 2003, White and Ernst 2003) argue for more attention to ecological objectives and better integration of ecological and social objectives. Other authors (TRB 2002b) suggest that transportation agencies become environmental stewards and address ecological issues earlier in the planning process through improved coordination, communication, and education among the agencies. The committee suggests that integration of environmental concerns with other social objectives may be facilitated in three areas: (1) developing new institutions and institutional arrangements, (2) developing new tools and methods, and (3) developing adaptive management. Each area is discussed in the subsequent sections.

### New Institutions and Arrangements

The committee's review and others (TRB 2002b) suggest that transportation agencies continue to expand their roles as environmental coordinators and stewards. This expansion can be viewed as filling "institutional gaps." Institutions are defined in this context as the set of norms and rules that people use to organize activities (Ostrom 1990). Institutional gaps occur when new problems arise that existing agency mandates do not address or when agencies recognize the need to interact in new ways with other agencies or stakeholders.

Expanding the existing mandate and tasks of transportation agencies toward environmental stewardship (Executive Order 13274, September 2002) has begun. It is a bureaucratic solution, the assumption being that governmental agencies can and will evolve and adapt to the new duties. There are a number of obstacles in attempting these changes, and the history of such attempts indicates mixed success. Recent attempts to reform agencies at the state level (such as water management districts in Florida) suggest that it is a long and difficult process (Light et al. 1995). The Corps attempted a similar reformation (Shallat 1994).

Another solution to the institutional gap is the creation of new institutions referred to as meshing organizations (Haas 1990), which would provide bridges among agencies and stakeholders that have existing relationships. These organizations may be formal or informal. One example of a formal arrangement is the tripartite model set up to manage resource issues in the Grand Canyon. Arising from an National Research Council (NRC) report (1996), three groups (the Grand Canyon Research and Monitoring Center, the Adaptive Management Work Group, and external review groups) meshed together to form the Adaptive Management Program. Forming a meshing group is one way to integrate technical and social concerns.

Other institutional forms exist. They include a type of ombudsman, in which the authority to resolve and reconcile technical and social issues is given to one individual. A modification of that approach could create a panel or council to resolve issues as they arise in the planning and assessment process. Dispute resolution mechanisms could also be considered to fill institutional gaps.

These types of institutions do not replace legal or scientific institutions that are in use. The new institutional forms are useful in sorting through assessments that do not require the rigor of a scientific approach to generate actions or plausible sets of actions. They also can be effective at reconciling or working around worldviews or mental models, which are often unspoken constructs that influence the way that environmental effects are perceived and managed. For example, the worldview of a real estate developer who works in downtown Atlanta might be very different from that of a conservationist who works to preserve biodiversity. There is a tension between the environmentally "negative" effects of road construction and use and the socially or economically "positive" effects of roads. Forums are needed for expression of many perspectives, so that differences can be recognized and resolutions negotiated.

Other institutional configurations, such as free markets, or some versions of cost-benefit approaches exist but are unlikely to resolve the complexities outlined above. Much work over the past decade or so in the field of ecological economics has attempted to quantify the value of ecosystem goods and services. Daly (1996), Costanza et al. (1997), Odum (1996), and others have created a variety of methods to place dollar values on the natural capital of ecosystems. Their work has undoubtedly created scholarly discourse and exposed the importance of the ecosystem goods and services that underpin but are ignored by economic institutions. Yet, few working examples that demonstrate the utility of an environmental cost-benefit approach to decision making about the environment are persuasive.

One or all of these proposals for new or alternative institutional arrangements could be attempted. In the wide array of governments and stakeholders across the United States, many (and even other new) combinations could be tested. In establishing these institutions, design principles developed by Ostrom (1990), who has studied international cooperative institutions that are organized and governed by the users of common-pool resources (for example, fisheries, groundwater basins, and irrigation systems) can be applied. The author's principles include clearly defined boundaries, provision and appropriation rules, collective choice rules, monitoring, gradual sanctioning, conflict resolution mechanisms, external recognition, and layers of nested enterprise. Her principles are useful guidelines for the design of new institutions that promote successful cooperation and achievement of social goals.

## New Tools and Methods

### Technological Developments

Advances in methods and technology for establishing and managing roads have reduced and will continue to reduce some of the ecological effects of roads. For example, wildlife underpasses can successfully retain the connection among populations of animals even when a road bisects their territory. Agencies such as the Natural Resource Conservation Service now supports nurseries in 18 locales for growing seed of native plants, which can thrive in the stressful roadside conditions. Even so, entire regions do not have such native plant nurseries and still heavily use a mix of largely European plants, which have been planted along roads for decades. Other developments have attempted to improve environmental benefits associated with roads:

- Innovations in building materials (for example, porous pavement), design strategies (for example, for bridges and culverts), and stormwater strategies are being developed and implemented to mitigate environmental impacts of roads.

- Since 1970, 44 states and Puerto Rico have constructed over 1,600 miles of noise barriers.
- Highway developers are among the largest recyclers in America through remilling pavements, using fly ash in concrete, and using crumb rubber as a component of road surfaces.
- Over the past 11 years, $4.9 billion in enhancement projects, such as bike paths and the preservation of historic bridges and train stations, has been spent in more than 14,000 communities (AASHTO 2002a).

Environmental effects of roads are greatly influenced by the surrounding socioeconomic conditions. Socioeconomic constraints and opportunities are often expressed in the use and management of roadsides. As an example of land-use constraints, roads in U.S. agricultural areas zigzag around fields. As human populations grow and change in distribution and movement patterns, roads are no longer designed to avoid fields and become linear features on the landscape. Furthermore, anticipated climate changes are likely to affect where people want to travel and where roads can be built (for example, seaside roads may be under water because of rising sea levels).

Technological change might alter the effects of roads on the environment. For example, air conditioning has changed where people want to live, and the Southeast and Southwest regions of the United States have grown rapidly. Telecommuting can reduce traffic volume. Although hydrogen-based fuels introduce new environmental problems, they would also reduce water, air, and noise pollution (Ludwig et al. 1993). Hence, changes in knowledge can affect how environmental effects, as well as steps taken to mitigate them, are viewed.

Several sites in the city of Seattle, Washington, provide examples of environmental improvements that can be implemented:

- The overpasses on Mercer Island, known as the 1-90 LIDS, are noted for their attractive dense vegetation.
- Freeway Park in downtown Seattle was designed to provide a bridge over Interstate 5 and connect adjoining neighborhoods (Wright 1989). Its waterfall masks traffic noise, and the gorge provides a quiet escape from city life. Using shallow-rooted trees and plants resistant to air pollution ensured the success of the park.
- Street edge alternatives (SEA) is a street design that uses special grading, soils, plants, and layout combined with traditional drainage

infrastructure to reduce traffic flow and water runoff in one block of a city neighborhood. The aesthetic benefits of this design increased property values, but it is unclear whether the increase makes up for the high cost of implementing a SEA.

For the most part, however, Seattle is a typical large city with all the environmental problems that accompany roads. The commitment to improved roadsides in spot locations probably arises from an interest in the environment being aesthetically appealing and in the federal listing of salmon as an endangered species; the effects of runoff on fish are a major environmental concern.

## New Conceptual Approaches

Roads can be envisioned as a network because they form a set of interconnections. One of the origins of the mathematical theory of networks deals with the transportation problem of moving goods and services across roads. Solutions to that and other optimization problems resulted in a body of knowledge on how goods can most efficiently move across interconnections. The recent application of these ideas to road systems has led to the formation of basic principles for interpreting how road networks fit into an ecological landscape (Forman et al. 2003):

- The arrangement of land uses around roads determines the structure of the road network.
- Network arrangement and traffic flow affect the delivery of good and services.
- The ecological landscape in which road networks are set can be influenced by road density, network structure, traffic flow, and patterns of the ecological systems in the landscape.
- The landscape patterns around a road network can have intended and unintended effects on roads and traffic. The design of the road network can be made ecologically sensitive.

Although the first two factors have been the focus of most road design, the last two factors call for consideration of the unintended ecological effects of roads.

The spatial attributes of the road network include the number and layout of roads. Road size is important because it affects traffic flow and road footprint (surface space). Road density (the lane length of roads per unit area of land) is perhaps the simplest measure of road structure. Road density can influence such diverse ecological features as how far large animals move across the landscape (van Dyke et al. 1986, Mech et al. 1988) and how fast water moves into streams (Jones and Grant 1996). Hence, attributes of road networks can serve as a means to quantify the human impact on the environment.

## Adaptive Management

This report and others (Forman et al. 2003) suggest that although much has been learned about the interactions between roads and ecosystems, much more is required. In addition to peer-reviewed published studies from academic and technical centers and summaries and syntheses by professional road groups, the committee recommends the development of other learning methods—that is, some retrospective analysis to evaluate efficacy of actions taken to avoid or mitigate ecological effects. In a sense, the approaches being applied are not solutions but rather best guesses at solutions. For example, when over- or underpasses are constructed for animal migration, there may be uncertainty about animal use, what kinds of animals will use or avoid them, and unexpected results (such as increased predation risks). Often, solutions applied in particular projects are, in effect, pilot projects to see whether and how a problem can be addressed. As noted elsewhere in the report, there is insufficient monitoring, follow-up, and reevaluation, the result being that the opportunity to learn from experience is lessened. Learning opportunities are also available from the management activities that are done throughout the country (see Chapter 4).

To evaluate construction or management actions, new sets of data, types of institutions, and ways of evaluating actions need to be developed. The need for a robust set of ecological indicators is central to this activity.

## Examining Ecological Indicators

Ecological indicators are meant to quantify the magnitude of stress, degree of exposure to stresses, and degree of ecological response to exposures (Hunsaker and Carpenter 1990, Suter 1993). They provide a simple and efficient method to examine the ecological composition, structure, and functioning of complex ecological systems (Karr 1981). Several ecological indicators have been proposed to measure or monitor ecological effects, and some of these indicators are applicable to road effects. Reports of the NRC (2000b), the Heinz Center (2002), and EPA (2003c) focus on indicators at the national level. The EPA framework (EPA/SAB 2002) builds on the Heinz Center report to form a checklist of factors to be considered for measurement. The journal Ecological Indicators and many scientific papers mark the growth of this field. Ecological Indicators for the Nation (NRC 2000b) evaluates indicators meant to track such factors as how land is used and the status of wildlife and other natural resources at a national level. These measures are meant to identify potential national environmental problems and evaluate the effectiveness of protective regulations and policies. The recommended indicators fall into three categories:

- **Ecosystem status.** Indicators include the types and extent of the nation's major land cover, such as wetlands, forests, and deserts.
- **Ecological capital.** The capacity of an ecological system to maintain itself as determined by such indicators as total species, native species diversity, nutrient runoff, and soil quality.
- **Ecosystem productivity.** Ecological indicators of a system's ability to produce oxygen and capture and store energy to support life include production capacity, carbon storage, and oxygen content in rivers, streams, and coastal areas.

The NRC (2000b) report calls for indicators to be credible, understandable, quantifiable, and broadly applicable (including roads). The data upon which the indicators depend should be clear and objective. The Heinz Center (2002) report presents a plan for regular reporting on the condition and use of the nation's lands, waters, and living resources. It identifies indicators for the nation's ecological systems, provides information on current conditions and past trends, and illustrates the many gaps in current data on key characteristics of ecological systems at the national level. Among these indicators are physical and chemical conditions (for example, nitrogen and phosphorus storages), changes in land use (for example, aerial extent of forests and grasslands), and listing and extent of nonnative plants and animals. The EPA (2003c) report on the environment focuses on how human activities affect human health and the environment at the national level. It quantifies trends in five areas:

- **Human health.** Changes in diseases, human exposure to environmental pollutants, and diseases that might be related to environmental pollution.
- **Ecological condition.** A perspective of natural resources, stressors on those resources, and potential for future sustainability.
- **Clean air.** Effects of indoor and outdoor air quality on human health and ecological systems.
- **Pure water.** Drinking water, use of recreational water, and condition of water resources and the living systems that they sustain.
- **Better protected land.** Land use and activities that affect the condition of the landscape, including agricultural practices, pest management, waste management, emergency response and preparedness, and recycling.

In summary, these three reports form a body of work on indicators at the national level that can be used to guide understanding of how best to measure ecological effects of roads at the scale of the entire United States. However, many of the effects of roads occur on local or intermediate scales. The use of ecological indicators relies on the assumption that the presence or absence of these indicators reflects changes occurring at various levels in the ecological hierarchy, from genes to species and ultimately to entire regions and the nation (Noon et al. 1999). The ecological system must be viewed as a moving target (Walters and Holling 1990), the system variables changing slowly and not stabilizing for a long period of time.

The EPA/SAB (2002) *Framework for Assessing and Reporting on Ecological Condition* provides a checklist that can be used to determine whether key ecological attributes are being considered at any scale of measure. It also provides a scheme for organizing the hundreds of proposed indicators. The major thematic areas of the framework are landscape condition, biotic condition, chemical and physical characteristics, ecological processes, hydrology and geomorphology, and disturbance regimes. Because these themes are independent of spatial scale, they formed the basis for the approach adopted in this report.

Several concerns constrain the use of ecological indicators as a resource management tool (Landres et al. 1988, Kelly and Harwell 1990, Noss 1990, Kremen 1992, Cairns et al. 1993, Mills et al. 1993, Noss and Cooperrider 1994, Gurney et al. 1995, Simberloff 1997). Dale and Beyeler (2001) summarized these concerns in three categories:

• Monitoring programs often depend on a small number of indicators and thus fail to consider the full complexity of the ecological system.

• Choice of ecological indicators is often confounded in management programs that have vague long-term goals and objectives.

• Management and monitoring programs often lack scientific rigor because of their failure to use a defined protocol for identifying ecological indicators.

Yet, selection of appropriate indicators is key to any planning, monitoring, or management program. Identifying criteria for using indicators for analyzing and addressing road effects on ecological systems might be the best way to determine the indicators most appropriate for roads effects. Cairns et al. (1993) pointed out that ecological indicators can be used to assess the condition of the environment, to monitor trends in condition over time, to provide an early warning signal of changes in the environment, and to diagnose the cause of an environmental problem. The purpose of using ecological indicators influences the choice of which indicator is most appropriate (Slocombe 1998). Several criteria should be used to evaluate indicators. These criteria are the significance of ecological changes measured by indicators, the basis of the ecological understanding of indicators, the reliability of indicators, the quality of the indicator data, and the costs and benefits (NRC 2000b). Criteria listed for indicators are drawn from Dale and Beyeler (2001):

• **The indicator is easily measured**. "The indicator should be straightforward and relatively inexpensive to measure. The metric needs to be easy to understand, simple to apply, and provide information to managers and policymakers that is relevant, scientifically sound, easily documented and cost-effective."

• **The indicator is sensitive to stresses on the system**. "While some indicators may respond to all dramatic changes in the system, the most useful indicator is one that displays high sensitivity to a particular and, perhaps, subtle stress, thereby serving as an early indicator of reduced system integrity."

• **The indicator responds to stress in a predictable manner**. "The indicator response should be unambiguous and predictable even if the indicator responds to the stress by a gradual change.... Ideally there is some threshold response level at which the observable response occurs before the level of concern."

- **The indicator is anticipatory (signifies an impending change in the ecological system)**. "Change in the indicator should be measurable before substantial change in ecological system integrity occurs."
- **The indicator predicts changes that can be averted by management actions**. "The value of the indicator depends on its relationship to possible changes in management actions."
- **The indicator has a known response to natural disturbances, anthropogenic stresses, and changes over time**. "The indicator should have a well-documented reaction to both natural disturbance and to anthropogenic stresses in the system. This criterion would pertain to conditions that have been extensively studied and have a clearly established pattern of response."
- **The indicator has low variability in response**. "Indicators that have a small range in response to particular stresses allow for changes in the response value to be better distinguished from background variability."
- **The indicator is integrative**. "The full suite of indicators provides a measure of coverage of the key gradients across the ecological systems (e.g., soils, vegetation types, temperature, etc.)."

Furthermore, information proposed to evaluate candidate indicators (see Andreasen et al. 2001) might be useful in evaluating the suite selected for a particular condition.

The spatial scale of the indicator affects both the availability of relevant data and the interpretation. For example, the Heinz Center (2002) report focuses on indicators at the national level and largely demonstrates the absence of information at this scale. The committee's analysis of data on ecological effects of roads (see Table 3-2) also reveals the paucity of data at the national scale. However, local and regional (or intermediate) scale information is more readily available and more interpretable in terms of analyzing road effects. Therefore, the committee recommends that projects designed to manage and monitor road effects should include indicators specific to the scale and concerns about the potential effects. After the goals and criteria for indictors for a specific project have been identified, the EPA/SAB (2002) framework can be used as a checklist to determine which indicators might be most useful. Perhaps the initiation of a few pilot programs that cover a wide range of spatial and temporal scales that engage relevant expertise on indicators (such as the NRC or the Heinz Center) would be a way to improve understanding of the development and application of key indicators, such as soil nutrient content, size of stream buffers, and distribution of nonnative pest organisms (plants and animals).

To date, attributes of road networks are not often considered in lists of ecological indicators, but road density and the spatial arrangement of new roads or roads that are removed strongly influence the ecological implications of road networks.

## SUMMARY

In evaluating ecological effects of roads, two areas of integration were identified. The first involves integration across scales and complexities of ecological systems. The second involves integration across multiple societal goals associated with roads.

Ecological systems cover wide ranges of spatial and temporal scales. The cross-scale structure and dynamics are not always amenable to simple scaling approaches and are the source of unexpected behavior. Some road effects confine the scale of ecological effects, and other effects can increase or decrease the spatial or temporal extent of ecological processes.

The complexity and cross-scale interactions in ecological systems generate problems for assessment and planning. Multiple assessments must be developed, each at different spatial and temporal scales to address key ecosystem processes and structures. Ecological concerns should be included early in the assessment and planning processes, which are context sensitive. Some components of the environment are incompatible with the existence of roads. Although great progress has been made in understanding and mitigating road effects, much more is needed. The development of a broader set of robust ecological indicators and learning-based institutions will help to facilitate understanding of the complexities in ecological systems.

Complexities that arise in the dynamics of social structures used to assess, plan, construct, and manage road systems must be addressed. Better integration of the social institutions will probably require the development of new relationships among the existing institutions. Transportation agencies have an opportunity to play a key role in meshing and integrating planning and management of environmental issues. New types of institutions are needed to address the mix of socioeconomic and ecological concerns. Enhanced collaboration can be generated by new kinds of rules and groups for interactions among agencies and stakeholders.

# 8

# Conclusions and Recommendations

On the basis of its investigations and deliberations, the committee reached the following conclusions and recommendations.

**CONCLUSION:** Most road projects today involve modifications to existing roadways, and the planning, operation, and maintenance of such projects often are opportunities for improving ecological conditions. A growing body of information describes such practices for improving aquatic and terrestrial habitats.

**Recommendation:** *The many opportunities that arise for mitigating or reducing adverse environmental impacts in modifications and repairs to existing roads should not be overlooked. Environmental considerations should be included when plans are made to repair or modify existing roads, as well as when plans are made to build new roads.*

**CONCLUSION:** Planning boundaries for roads and assessing associated environmental effects are often based on socioeconomic considerations, resulting in a mismatch between planning scales and spatial scales at which ecological systems operate. In part, this mismatch results because there are few legal incentives or disincentives to consider environmental effects beyond political jurisdictions, and thus decision making remains primarily local. The ecological effects of roads are typically much larger than the road itself, and they often extend beyond regional planning domains.

Scientific literature on ecological effects of roads generally addresses local-to-intermediate scales, and many of those effects are well documented. However, there are few integrative or large-scale studies. Sometimes the appropriate spatial scale for ecological research is not known in advance, and in that case, some ecological effects of roads may go undetected if an inappropriate scale is chosen. Few studies have addressed the complex nature of the ecological effects of roads, and the studies that have done so were often based on small sampling periods and insufficient sampling of the range of variability in ecological systems.

**Recommendation:** *Research on the ecological effects of roads should be multiscale and designed with reference to ecological conditions and appropriate levels of organization (such as genetics, species and populations, communities, and ecological systems).*

**Recommendation:** *Additional research is needed on the long-term and large-scale ecological effects of roads (such as watersheds, eco-regions, and species' ranges). Research should focus on increasing the understanding of cross-scale interactions.*

**Recommendation:** *More opportunities should be created to integrate research on road ecology into long-term ecological studies by using long-term ecological research sites and considering the need for new ones.*

**Recommendation:** *Ecological assessments for transportation projects should be conducted at different time scales to address impacts on key ecological system processes and structures. A broader set of robust ecological indicators should be developed to evaluate long-term and broad-scale changes in ecological conditions.*

**CONCLUSION:** The assessment of the cumulative impacts of road construction and use is seldom adequate. Although many laws, regulations, and policies require some consideration of ecological effects of transportation activities, such as road construction, the legal structure leaves substantial gaps in the requirements. Impacts on certain resources are typically authorized through permits. Permitting programs usually consider only direct impacts of road construction and use on a protected resource, even though indirect or cumulative effects can be substantial (for example, effects on food web components). The incremental effects of many impacts over time could be significant to such resources as wetlands or wildlife.

**Recommendation:** *More attention should be devoted to predicting, planning, monitoring, and assessing the cumulative impacts of roads. In some cases, the appropriate spatial scale for the assessment will cross state boundaries, and especially in those cases, collaboration and cooperation among state agencies would be helpful.*

**CONCLUSION:** The methods and data used for environmental assessment are insufficient to meet the objectives of rapid assessment, and there are no national standards for data collection. However, tools for in situ monitoring, remotely sensed monitoring, data compilation, analysis, and modeling are continually being improved, and because of advances in computer technology, practitioners have quick access to the tools. The new and improved tools now allow for substantial improvements in environmental assessment.

**Recommendation:** *Improvements are needed in assessment methods and data, including spatially explicit models. A checklist addressing potential impacts should be adapted that can be used for rapid assessment. Such a checklist would focus attention on places and issues of greatest concern. A national effort is needed to develop standards for data collection. A set of rapid screening and assessment methods for environmental impacts of transportation and a national ecological database based on the geographic information system (GIS) and supported by multiple agencies should be developed and maintained for ecological effects assessment and ecological system management across all local, state, and national transportation, regulatory, and resource agencies. Standard GIS data on road networks (for example, TIGER) could be interfaced with data models (for example, UNETRANS) to further advance the assessment of ecological impacts of roads.*

**Recommendation:** *The committee recommends a new conceptual framework for improving integration of ecological considerations into transportation planning. A key element of this framework is the integration of ecological goals and performance indicators with transportation goals and performance indicators.*

**Recommendation:** *Improved models and modeling approaches should be developed not only to predict how roads will affect environmental conditions but also to improve communication in the technical community, to resolve alternative hypotheses, to highlight and evaluate* *data and environmental monitoring, and to provide guidance for future environmental management.*

**CONCLUSION:** With the exception of certain legally specified ecological resources, such as endangered or threatened species and protected wetlands, there is no social or scientific consensus on which ecological resources affected by roads should be given priority attention. In addition, current planning assessments that focus on transportation needs rarely integrate other land-management objectives in their assessments.

**Recommendation:** *A process should be established to identify and evaluate ecological assets that warrant greater protection. This process would require consideration not only of the scientific questions but also of the socioeconomic issues. The Federal Highway Administration should consider amending its technical guidance, policies, and regulations based on the results of such studies.*

**CONCLUSION:** The state transportation project system offers the opportunity to consider ecological concerns at early planning stages. However, planning at spatial and temporal scales larger than those currently considered, generally does not address ecological concerns until later in a project's development.

**Recommendation:** *Environmental concerns should be integrated into transportation planning early in the planning process, and larger spatial scales and longer time horizons should be considered. Adding these elements would help to streamline the planning process. Metropolitan planning organizations and state departments of transportation should conduct first-level screenings for potential environmental effects before the development of a transportation improvement plan. Transportation planners should consider resource-management plans and other agencies' (such as the U.S. Corps of Engineers, U.S. Environmental Protection Agency, U.S. Fish and Wildlife Service, and National Park Service) environmental plans and policies as part of the planning process. Other agencies should incorporate transportation forecasting into resource planning.*

**CONCLUSION:** Elements of the transportation system, including the types of vehicles and their fuels, will continue to evolve. Changes in traffic volume and road capacity, mostly through widening of roads rather than construction of new corridors, have smaller but nevertheless important ecological effects compared with the creation of new, paved roads.

**Recommendation:**  *Monitoring systems should be developed for the evaluation and assessment of environmental effects resulting from changes in the road system—for example, traffic volume, vehicle mix, structure modifications, and network adjustments. Data from monitoring could then be used to evaluate previous assessments and, over the long term, improve understanding of ecological impacts.*

**CONCLUSION:**  Much useful information from research on the ecological effects of roads is not widely available because it is not in the peer-reviewed literature.  For example, studies documenting the effects of roads on stream sedimentation have been reported in documents of state departments of transportation, the U.S. Army Corps of Engineers, and the World Bank.  Although much of this literature is available through bibliographic databases, it is not included in scientific abstracting services and may not be accessible to a broader research community. Also, the data needed to evaluate regulatory programs are not easily accessible or amenable to synthesis.  The data are typically contained in project-specific environmental impact statements, environmental assessments, records of decision, or permits (for example, wetlands permits), which are not easily available to the scientific community.

**Recommendation:**  *Studies on ecological effects of roads should be made more accessible through scientific abstracting services or through publication in peer-reviewed venues.  The Federal Highway Administration, in partnership with state and federal resource-management agencies, should develop environmental information and decision-support systems to make ecological information available in searchable databases.*

**CONCLUSION:**  Transportation agencies have been attempting to fill an institutional gap in ecological protection created by the multiple social and environmental issues that must be addressed at all phases of road development.  The gaps often occur when problems arise that are not covered by agency mandates or when agencies need to interact with other organizations in new ways.  Even when transportation agencies work toward environmental stewardship, they cannot always do the job alone.

**Recommendation:** *Transportation agencies should continue to expand beyond their historical roles as planners and engineers, increasing their roles as environmental coordinators and stewards. Transportation planners and natural-resource planners should collaborate to promote integrated planning at comparable scope and scale so that the efforts can support mutual objectives. This collaboration should include federal, state, and county resource-management agencies; nongovernmental organizations; and organizations and firms involved in road construction. Incentives, such as funding and technical support, should be provided to help planning agencies, resource agencies, nongovernmental groups, and the public to understand ecological structure and functioning across jurisdictions and to interact cooperatively.*

# References

AASHTO (American Association of State Highway and Transportation Officials). 1993. AASHTO Guide for Design of Pavement Structures, 1993. GDPS-4. Washington, DC: American Association of State Highway and Transportation Officials.

AASHTO (American Association of State Highway and Transportation Officials). 2002a. Transportation Invest in America, The Bottom Line [online]. Available: http://transportation.org/bottomline/bottomline2002.pdf [accessed Feb. 13, 2004].

AASHTO (American Association of State Highway and Transportation Officials). 2002b. Roadside Design Guide, 3rd Ed. Washington, DC: American Association of State and Highway Transportation Officials.

AASHTO (American Association of State Highway and Transportation Officials). 2003. Transportation Invest in America, Listening to America, Report on AASHO's TEA_21 Reauthorization Regional Forums [online]. Available: http://transportation.org/download/ListeningToAmerica.pdf [accessed Feb. 18, 2004].

Allen, T.F.H., and T.W. Hoekstra. 1992. Toward a Unified Ecology. New York: Columbia University Press.

Allen, T.F.H., and T.B. Starr. 1982. Hierarchy: Perspectives for Ecological Complexity. Chicago, IL: University of Chicago Press.

AMPO (Association of Metropolitan Planning Organization). 2004. Metropolitan Planning Organization Website [online]. Available: http://www.ampo.org/ [accessed Feb. 18, 2004].

Andreasen, J.K., R.V. O'Neill, R. Noss, and N.C. Slosser. 2001. Considerations for the development of a terrestrial index of ecological integrity. Ecological Indicators 1(1):21-35.

Angold, P.G. 1997. The impact of a road upon adjacent heathland vegetation: Effects on plant species composition. J. Appl. Ecol. 34(2):409-417.

Argetsinger, A. 2005. Vegas suburb may rise from the Tumbleweeds: Developers eye Arizona desert for spillover growth. Washington Post, P. A3, March 26, 2005.

Asplund, R., B.W. Mar, and J.F. Ferguson. 1982. Total suspended solids in highway runoff in Washington State. J. Environ. Eng.-ASCE 108(EE2): 391-404.

Bailey, R.G., P.E. Avers, T. King, and W.H. McNab, eds. 1994. Ecoregions and Subregions of the United States (map). Reston, VA: U.S. Geological Survey.

Baker, R.H. 1998. Are man-made barriers influencing mammalian speciations? J. Mammal. 79:370-371.

Ball, J.E., R. Jenks, and D. Aubourg. 1998. An assessment of the availability of pollutant constituents on road surfaces. Sci. Total Environ. 209(2-3):243-254.

Bank, F.G., C.L. Irwin, G.L. Evink, M.E. Gray, S. Hagood, J.R. Kinar, A. Levy, D. Paulson, B. Ruediger, R.M. Sauvajot, D.J. Scott, and P. White. 2002. Wildlife Habitat Connectivity across European Highways. FHWA-PL-02-011. Washington, DC: Office of International Programs, Federal Highway Administration, U.S. Department of Transportation [online]. Available: http://www.international.fhwa.dot.gov/Pdfs/wildlife_web.pdf [accessed Feb.18, 2004].

Barnett, J.L., R.A. How, and W.F. Humphreys. 1978. The use of habitat components by small mammals in eastern Australia. Aust. J. Ecol. 3:277–285.

Barrett, M.E., J.F. Malina, R.J. Charbeneau, and G.H. Ward. 1995. Characterization of Highway Runoff in the Austin, Texas Area. Technical Report CWRW 263. Austin, TX: Center for Research in Water Resources, the University of Texas at Austin.

Barrett, M.E., L.B. Irish, Jr., J.F. Malina, Jr., and R.J. Charbeneau. 1998. Characterization of highway runoff in Austin, Texas area. J. Environ. Eng. 124(2):131-137.

Bates, K.M., R.J. Barnard, B. Heiner, J.P. Klavas, and P.D. Powers. 2003. Design of Road Culverts for Fish Passage. Washington Department of Fish and Game, Olympia, WA [online]. Available: http://wdfw.wa.gov/hab/engineer/cm/culvert_manual_final.pdf [accessed April 14, 2005].

Beier, P. 1995. Disposal of juvenile cougars in fragmented habitat. J. Wildl. Manage. 59:228-237.

Bekker, G.J., and K.J. Canters. 1997. The continuing story of badgers and their tunnels. Pp. 344-353 in Habitat Fragmentation and Infrastructure: Proceedings of the International Conference "Habitat Fragmentation, Infrastructure and the Role of Ecological Engineering," 17-21 September, Maastricht, The Hague, The Netherlands, K. Canters, ed. Delft, The Netherlands: Ministry of Transport, Public Works and Water Management, Directorate-General for Public Works and Water Management, Road and Hydraulic Engineering Division.

Bekker, G.J., B. van den Hengel, H. Bohemen, and H. van der Sluijs. 1995. Nature over Wegen [Nature over Motorways]. Ministry of Transport, Public Works and Water Management, Delft, The Netherlands.

Bergquist, G., and C. Bergquist. 1999. Post-decision assessment. Pp. 285-316 in Tools to Aid Environmental Decision Making, V.H. Dale, and M.R. English, eds. New York: Springer.

Berkes, F., and C. Folke, eds. 1998. Linking Social and Ecological Systems: Management Practices and Social Mechanisms for Building Resilience. Cambridge, UK: Cambridge University Press. 459 pp.

Berkovitz, A. 2001. The marriage of safety and land-use planning: A fresh look at local roadways. Public Roads 65(2) [online]. Available: http://www.tfhrc.gov/pubrds/septoct01/marriage.htm [accessed June 2, 2004].

Blaesser, B.W., D.R. Mandelker, M.S. Giaimo, and J.B. McDaniel. 2003. Selected Studies in Transportation Law: Vol. 3. Environmental Law and Transportation. National Cooperative Highway Research Program, CD-ROM. Washington, DC: Transportation Research Board, The National Academies.

Bocca, B., F. Petrucci, A. Alimonti, and S. Caroli. 2003. Traffic-related platinum and rhodium concentrations in the atmosphere of Rome. J. Environ. Monitor. 5(4):563-568.

Boeke, K. 1957. Cosmic View: The Universe in Forty Jumps. New York: J. Day.

Boudet, C., D. Zmirou, and D. Poizeau. 2000. Fraction of PM2.5 personal exposure attributable to urban traffic: A modeling approach. Inhal. Toxicol. 12(Suppl.1): 41-53.

Brandenburg, D.M. 1996. Effects of Roads on Behavior and Survival of Black Bears in Coastal North Carolina. M.S. Thesis, University of Tennessee, Knoxville, TN.

Brinson, M.M. 1993. A Hydrogeomorphic Classification for Wetlands. Technical Report WRP-DE-4. U.S. Army Engineer Waterways Experiment Station, Vicksburg, MS [online]. Available: http://www.wes.army.mil/el/wetlands/pdfs/wrpde4.pdf [accessed Feb. 18, 2004].

Brinson, M.M. 1995. The HGM approach explained. National Wetlands Newsletter 17(6):7-13.

Brinson, M.M. 1996. Assessing wetland functions using HGM. National Wetlands Newsletter 18(1):10-16.

Brown, J.H., and M.V. Lomolino. 1998. Biogeography, 2nd Ed. Sunderland, MA: Sinauer Associates.

Buckeridge, D.L., R. Glazier, B.J. Harvey, M. Escobar, C. Amrhein, and J. Frank. 2002. Effect of motor vehicle emissions on respiratory health in an urban area. Environ. Health Perspect. 110(3):293-300.

Butler, R., P. Angelstam, P. Ekelund, and R. Schlaeffer. 2004. Dead wood threshold values for the three-toed woodpecker presence in boreal and sub-Alpine forest. Biol. Conserv. 119(3):305-318.

Buzas, K., and L. Somlyody. 1997. Impacts of road traffic on water quality. Per. Pol. Civil Eng. 41(2):95-106.

Cairns, J., P.V. McCormick, and B.R. Niederlehner. 1993. A proposed framework for developing indicators of ecosystem health. Hydrobiologia 236(1):1-44.

Carpenter, S.R., and K.L. Cottingham. 2002. Resilience and the restoration of lakes. Pp. 51-70 in Resilience and the Behavior of Large Scale Systems, L.H. Gunderson, and L. Pritchard Jr., eds. Washington, DC: Island Press.

Carr, L.W., and L. Fahrig. 2001. Effect of road traffic on two amphibian species of differing vagility. Conserv. Biol. 15(4):1071-1078.

Carr, M.H., P.D. Zwick, T. Hoctor, W. Harrel, A. Goethals, and M. Benedict. 1998. Using GIS for identifying interface between ecological greenways and roadway systems at the State and sub-state scales. Pp. 68-77 in Proceedings of the International Conference on Wildlife Ecology and Transportation (ICOWET), February 10-12, 1998, Fort Meyers, FL, G.L. Evink, P. Garrett, D. Zeigler, and J. Berry, eds. FL-ER-69-98. Washington, DC: U.S. Department of Transportation, Federal Highway Administration.

Carson, R. 1962. Silent Spring. Boston: Houghton Mifflin.

Carstens, K. 2003. Making a noise about environmental pollution. European Voice 9(38):19.

Chruszcz, B., A.P. Clevenger, K.E. Gunson, and M.L. Gibeau. 2003. Relationships among grizzly bears, highways, and habitat in the Banff-Bow Valley, Alberta, Canada. Can. J. Zool./Rev. Can. Zool. 81(8):1378-1391.

CIFOR (Center for International Forestry Research). 2004. MLA: Multidisciplinary Landscape Assessment, Center for International Forestry Research [online]. Available: http://www.cifor.cgiar.org/mla/ [accessed Oct. 5, 2004].

Clark, W.D., and J.R. Karr. 1979. Effect of highways on red-winged blackbird and horned lark populations. Wilson Bull. 91:143-145.

Clevenger, A.P., B. Chruszcz, and K. Gunson. 2001. Drainage culverts as habitat linkages and factors affecting passage by mammals. J. Appl. Ecol. 38(6):1340-1349.

Clevenger, A.P., J. Wierzchowski, B. Chruszcz, and K. Gunson. 2002a. GIS-generated, expert based models for identifying wildlife habitat linkages and planning mitigation passages. Conserv. Biol. 16(2):503-514.

Clevenger, A.P., B. Chruszcz, K. Gunson, and J. Wierzchowski. 2002b. Roads and Wildlife in the Canadian Rocky Mountain Parks - Movements, Mortality and Mitigation. Final Report prepared for Parks Canada, Banff, Alberta, Canada.

Clifford, M.J., R. Clarke, and S.B. Riffat. 1997. Local aspects of vehicular pollution. Atmos. Environ. 31(2):271-276.

Cohen, W.B., and S.N. Goward. 2004. Landsat's role in ecological applications of remote sensing. Bioscience 54(6):535-545.

Committee of Scientists. 1999. Sustaining the People's Lands: Recommendations for Stewardship of the National Forests and Grasslands into the Next Century, March 15, 1999. U.S. Department of Agriculture, Washington, DC [online]. Available: http://www.fs.fed.us/news/news_archived/science/cosfrnt.pdf [accessed Feb. 18, 2004].

Conrey, R.C., and L.S. Mills. 2001. Do highways fragment small mammal populations? Pp. 448-457 in Proceedings of the 2001 International Conference on Ecology and Transportation, Sept. 24-28, 2001, Keystone, CO, G.L. Evink, ed. Raleigh, NC: North Carolina State University, Center for Transportation and the Environment [online]. Available: http://itre.ncsu.edu/cte/icoet/downloads/Wildlife.pdf [accessed Feb. 24, 2004].

Cook, K.E., and P.M. Daggett. 1995. Highway Road Kill and Associated Issues of Safety and Impacts on Highway Ecotones. Task Force on Natural Resources, Transportation Research Board, National Research Council, Washington, DC.

Costanza, R., R. d'Arge, R. de Groot, S. Farber, M. Grasso, B. Hannon, K. Limburg, S. Naeem, R.V. O'Neill, J. Paruelo, R.G. Raskin, P. Sutton, and M. van den Belt. 1997. The value of the world's ecosystem services and natural capital. Nature 387(6630):253-260.

Daily, G.C., ed. 1997. Nature's Services: Societal Dependence on Natural, Ecosystems. Washington, DC: Island Press.

Dale, V.H. 2003. New directions in ecological modeling for resource management. Pp. 310-320 in Ecological Modeling for Resource Management, V.H. Dale, ed. New York: Springer.

Dale, V.H., and S.C. Beyeler. 2001. Challenges in the development and use of ecological indicators. Ecological Indictors 1(1):3-10.

Dale, V.H., and R.A. Haeuber, eds. 2001. Applying Ecological Principles to Land Management. New York: Springer.

Dale, V.H., R.V. O'Neill, F. Southworth, and M. Pedlowski. 1994. Modeling effects of land management in the Brazilian Amazonian settlement of Rondonia. Conserv. Biol. 8(1):196-206.

Dale, V.H., S. Brown, R.A. Haeuber, N.T. Hobbs, N. Huntly, R.J. Naiman, W.E. Riebsame, M.G. Turner, and T.J. Valone. 2000. Ecological principles and guidelines for managing the use of land. Ecol. Appl. 10(3):639-670.

Dale, V.H., S.C. Beyeler, and B. Jackson. 2002. Understory vegetation indicators of anthropogenic disturbance in longleaf pine forests at Fort Benning, Georgia, USA. Ecological Indicators 1(3):155-170.

Daly, H. 1996. Beyond Growth: The Economics of Sustainable Development. Boston: Beacon Press.

Damarad, T. 2003. Final Report of the COST Action 341" Habitat Fragmentation Due to Transport Infrastructure." European Centre for Nature Conservation, Ministry of Transport, Public Works and Water Management of the Netherlands, Road and Hydraulic Engineering Division.

Davis, S.M., and J.C. Ogden, eds. 1994. Everglades: The Ecosystem and its Restoration. Delray Beach, FL: St. Lucie Press.

Defenders of Wildlife. 2004. Habitat and Highways Campaign. TEA-21 Reauthorization: What's Brewing for Wildlife? [online]. Available: http://www.defenders.org/habitat/highways/ [accessed Feb. 17, 2004].

Delfino, R.J., H. Gong, W.S. Linn, Y. Hu, and E.D. Pellizzari. 2003. Respiratory symptoms and peak expiratory flow in children with asthma in relation to volatile organic compounds in exhaled breath and ambient air. J. Expo. Anal. Environ. Epidemiol. 13(5):348-363.

De Molenaar, J.G., D.A. Jonkers, and M.E. Sanders. 2000. Road Lighting and Nature III: Local Influence of Road Lighting on a Population of Gruttos [in Dutch]. Dienst Weg-en Waterbouwkunde/Instituut voor Bos-en Natuuronderzoek, Delft/Wageningen, The Netherlands.

De Molenaar, J.G., R.J.H.G. Henkens, C. ter Braak, C. van Duyne, G. Hoefsloot, and D.A. Jonkers. 2003. Road Illumination and Nature. IV. Effects of Road Lights on the Spatial Behaviour of Mammals. DWW Report No. DWW-2008-044. Alterra, Green World Research, Wageningen, NL [online]. Available: http://www.minvenw.nl/rws/dww/uitgaven/grutto/downloads/44EN_wegverlichting.pdf [accessed Feb. 9, 2004].

De Nevers, N. 2000. Air Pollution Control Engineering, 2nd Ed. Boston: McGraw-Hill.

DEQ (Virginia Department of Environmental Quality). 2004. Virginia Water Protection Permit Program, Wetlands, Surface Water, and Surface Water, Withdrawals/ Impoundments [online]. Available: http://www.deq.state.va.us/wetlands/ [accessed Feb. 17, 2004].

Dongarra, G., G. Sabatino, M. Triscari, and D. Varrica. 2003. The effects of anthropogenic particulate emissions on roadway dust and Nerium oleander leaves in Messina (Sicily, Italy). J. Environ. Monitor. 5(5):766-773.

DOT (U.S. Department of Transportation). 2003. Transportation Statistics Annual Report, October 2003. Bureau of Transportation Statistics, U.S. Department of Transportation, Washington, DC [online]. Available: http://www.bts.gov/publications/transportation_statistics_annual_report/2003/ [accessed June 2, 2004].

Durbrow, B.R., N.B. Burns, J.R. Richardson, and C.W. Berish. 2001. Southeastern ecological framework: A planning tool for managing ecosystem integrity. Pp. 352-357 in Proceedings of the 2001 Georgia Water Resources Conference, K.J. Hatcher, ed. Athens, GA: Institute of Ecology, University of Georgia.

Eaglin, G.S., and W.A. Hubert. 1993. Effects of logging and roads on substrate and trout in streams of the Medicine Bow National Forest, Wyoming. N. Am. J. Fish. Manage. 13(4):844-846.

Ellis, J.B., D.M. Revitt, and N. Llewellyn. 1997. Transport and the environment: Effects of organic pollutants on water quality. Water Environ. Manage. J. 11(3)170-177.

Environment Canada. 2003. Road Salts. Priority Substance List –Assessment Reports, Series 2. PSL 63. Environment Canada [online]. Available: http://www.ec.gc.ca/substances/ese/eng/psap/final/roadsalts.cfm [accessed May 26, 2004].

EPA (U.S. Environmental Protection Agency). 2003a. Acid Rain. Office of Air and Radiation, U.S. Environmental Protection Agency [online]. Available: http://www.epa.gov/air/acidrain/ [accessed May 26, 2004].

EPA (U.S. Environmental Protection Agency). 2003b. National Estuary Program. Office of Wetlands, Oceans and Watersheds, U.S. Environmental Protection Agency [online]. Available: http://www.epa.gov/owow/estuaries/ [accessed Feb. 17, 2004].

EPA (U.S. Environmental Protection Agency). 2003c. EPA's Draft Report on the Environment. EPA 600-R-03-050. Office of Research and Development, Office of Environmental Information, U.S. Environmental Protection Agency, Washington, DC [online]. Available: http://www.epa.gov/indicators/roe/html/roePDF.htm [accessed Feb. 11, 2004].

EPA (U.S. Environmental Protection Agency). 2004. Mitigation/Mitigation Banking. Policy and Technical Guidance Documents. Office of Wetlands, Office of Wetlands, Oceans and Watersheds, U.S. Environmental Protection Agency [online]. Available: http://www.epa.gov/owow/wetlands/guidance/index.html#mitigation [accessed May 26, 2004].

EPA/SAB (U.S. Environmental Protection Agency Science Advisory Board). 2002. A Framework for Assessing and Reporting on Ecological Condition: An SAB Report. EPA-SAB- EPEC-02-009. U.S. Environmental Protection Agency Science Advisory Board, Washington, DC [online]. Available: www.epa.gov/sab/pdf/epec02009.pdf [accessed Feb. 11, 2004].

Evink, G.L. 2002. Interaction Between Roadways and Wildlife Ecology: A Synthesis of Highway Practice. National Cooperative Highway Research Program Report NCHRP Synthesis 305. Washington, DC: Transportation Research Board, The National Academies.

Fahrig, L., J.H. Pedlar, S.E. Pope, P.D. Taylor, and J.F. Wegner. 1995. Effect of road traffic on amphibian density. Biol. Conserv. 73(3):177-182.

Fakayode, S.O., and B.I. Olu-Owolabi. 2003. Heavy metal contamination of roadside topsoli in Osogbo, Nigeria: Its relationship to traffic density and proximity to highways. Environ. Geol. 44(2):150-157.

Ferris, C.R. 1979. Effects of Interstate 95 on breeding birds in northern Maine. J. Wildl. Manage. 43(2):421-427.

FHWA (Federal Highway Administration). 1979. America's Highways 1776-1976: A History of the Federal Aid Program. Washington, DC: U.S. Department of Transportation.

FHWA (Federal Highway Administration). 1981. Constituents of Highway Runoff. FHWA/RD-81/042- FHWA/RD-81/046. Washington, DC: Federal Highway Administration.

FHWA (Federal Highway Administration). 1987. Pp. 20-41 in Technical Advisory Guidance for Preparing and Processing Environmental and Section 4(F) Documents, T 6640.8A October 30, 1987. Federal Highway Administration, U.S. Department of Transportation [online]. Available: http://www.fhwa.dot.gov/legsregs/directives/techadvs/t664008a.htm [accessed Feb. 17, 2004].

FHWA (Federal Highway Administration). 1995. Complete integration of environmental concerns. P. 3 in Environmental Policy Statement 1994: A Framework to Strengthen the Linkage Between Environmental and Highway Policy. FHWA-PD-95-006(10M)E/HEP-32/R 4-95 (5M) EW. Washington, DC: Federal Highway Administration [online]. Available: http://www.fhwa.dot.gov/environment/eps/pt4.htm [accessed June 3, 2004].

FHWA (Federal Highway Administration). 2001a. TEA-21, Moving Americans into the 21st Century. Federal Highway Administration, U.S. Department of Transportation [online]. Available: http://www.fhwa.dot.gov/tea21/index.htm [accessed Feb. 11, 2004].

FHWA (Federal Highway Administration). 2001b. Our Nation's Highways 2000. Washington, DC: Federal Highway Administration [online]. Available: http://www.fhwa.dot.gov/ohim/onh00/our_ntns_hwys.pdf [accessed Feb. 13, 2004].

FHWA (Federal Highway Administration). 2002a. Highway Statistics 2001. Federal Highway Administration, U.S. Department of Transportation [online]. Available: http://www.fhwa.dot.gov/ohim/hs01/ [accessed Feb. 9, 2004].

FHWA (Federal Highway Administration). 2002b. Operational Performance. Chapter 4 in Status of the Nation's Highways, Bridges, and Transit: 2002 Conditions and Performance Report. Report to the Congress [online]. Available: http://www.fhwa.dot.gov/policy/2002cpr/cp02_pdf.htm [accessed Sept. 8, 2004].

FHWA (Federal Highway Administration). 2002c. Environmental Justice. Federal Highway Administration, U.S. Department of Transportation [online]. Available: http://www.fhwa.dot.gov/environment/ej2.htm [accessed June 1, 2004].

FHWA (Federal Highway Administration). 2003a. Keeping It Simple: Easy Ways to Help Wildlife Along Roads. Sonar "Startle" Devices Keep Herring Away From Blasting [online]. Available: http://www.fhwa.dot.gov/environment/wildlifeprotection/index.cfm?fuseaction=home.viewArticle&articleID=68 [accessed March 9, 2005].

FHWA (Federal Highway Administration). 2003b. Information: Exemplary Ecosystem Initiatives Criteria. Memorandum from James M. Shrouds, Di-

rector, Office of Natural and Human Environment, to Division Administrators, Federal Lands Highway Division Engineers, Federal Highway
Administration, U.S. Department of Transportation. October 6, 2003
[online]. Available: http://www.fhwa.dot.gov/environment/ecoinitm.htm
[accessed Feb. 17, 2004].

FHWA (Federal Highway Administration). 2003c. Critter Crossings, Linking
Habitats and Reducing Roadkill. Federal Highway Administration, U.S.
Department of Transportation [online]. Available: http://www.fhwa.dot.
gov/environment/wildlifecrossings/ [accessed Feb. 17, 2004].

FHWA (Federal Highway Administration). 2003d. Ecosystem Management
Guidance and Agreements. Federal Highway Administration, U.S. Department of Transportation [online]. Available: http://www.fhwa.dot.
gov/environment/endg_abs.htm [accessed Feb. 17, 2004].

FHWA (Federal Highway Administration). 2003e. Federal Guidance on the
Use of the TEA-21 Preference for Mitigation Banking to Fulfill Mitigation
Requirements under Section 404 of the Clean Water Act, July 11, 2003.
Federal Highway Administration, U.S. Department of Transportation
[online]. Available: http://www.fhwa.dot.gov/environment/wetland/
tea21bnk.htm [accessed Feb. 18, 2004].

FHWA (Federal Highway Administration). 2003f. Evaluating the Performance
of Environmental Streamlining: Development of a NEPA Baseline for
Measuring Continuous Performance. Prepared by The Louis Berger
Group, Inc., for the Federal Highway Administration [online]. Available:
http://environment.fhwa.dot.gov/strmlng/baseline/ [accessed June 2,
2004].

FHWA (Federal Highway Administration). 2004a. Environment Guidebook
Website. Federal Highway Administration, U.S. Department of Transportation [online]. Available: http://environment.fhwa.dot.gov/guidebook/
[accessed Feb. 17, 2004].

FHWA (Federal Highway Transportation). 2004b. Planning. Planning, Environment and Reality, Federal Highway Administration, U.S. Department
of Transportation [online]. Available: http://www.fhwa.dot.gov/
environment/index.htm [accessed Feb. 19, 2004].

FHWA (Federal Highway Administration). 2004c. Overview of Vital Few
Goal (VFG). FHWA's Vital Few Goal - Environmental Stewardship and
Streamlining. Federal Highway Administration, U.S. Department of
Transportation [online]. Available: http://environment.fhwa.dot.gov/
strmlng/vfovervw.htm [accessed June 10, 2004].

FHWA/FTA (Federal Highway Administration and Federal Transit Administration). 2005. Integration of Planning and NEPA Processes. Memorandum
to Cindy Burbank, Associate Administrator, Office of Planning, Environment and Realty, Federal Highway Administration, and David D. Voz-
zolo, Deputy Associate Administrator, Office of Planning and Environment, Federal Transit Administration, from D.J. Gribbin, Chief Counsel,

FHWA, and Judith S. Kaleta, Acting Chief Counsel, FTA. February 22, 2005 [online]. Available: http://environment.fhwa.dot.gov/strmlng/integmemo.htm [accessed May 11, 2005].

Findlay, C.S., and J. Bourdages. 2000. Response time of wetland biodiversity to road construction on adjacent lands. Conserv. Biol. 14(1):86-91.

Findlay, C.S., and J. Houlahan. 1997. Anthropogenic correlates of species richness in southeastern Ontario wetlands. Conserv. Biol. 11(4):1000-1009.

Fleck, A.M., M.J. Lacki, and J. Sutherland. 1988. Response by white birch (Betula papyrifera) to road salt applications at Cascade Lakes, New York. J. Environ. Manage. 27(4):369-377.

Flores-Rodriguez, J., A.L. Bussy, and D.R. Thevenot. 1994. Toxic metals in urban runoff: Physico-chemical mobility assessment using speciation schemes. Water Sci. Technol. 29(1-2):83-94.

Foppen, R., and R. Reijnen. 1994. The effects of car traffic on breeding bird populations in woodland. II. Breeding dispersal of male willow warblers (Phylloscopus trochilus) in relation to the proximity of a highway. J. Appl. Ecol. 31(1):95-101.

Foresman, K.R. 2003. Small mammal use of modified culverts on the Lolo South Project of Western Montana – an update. Pp. 342-343 in International Conference on Ecology and Transportation, Making Connections: 2003 Proceedings: August 24-29, 2003, Lake Placid, New York, C.L. Irwin, P. Garrett, and K.P. McDermott, eds. Raleigh, NC: Center for Transportation and the Environment, North Carolina State University [online]. Available: http://www.icoet.net/downloads/03Monitoringof Structures.pdf [accessed March 10, 2005].

Forman, R.T.T. 1987. The ethics of isolation, the spread of disturbance, and landscape ecology. Pp. 213-229 in Landscape Heterogeneity and Disturbance, M.G. Turner, ed. New York: Springer.

Forman, R.T.T. 2000. Estimate of the area affected ecologically by the road system in the United States. Conserv. Biol. 14(1):31-35.

Forman, R.T.T., and L.E. Alexander. 1998. Roads and their major ecological effects. Annu. Rev. Ecol. Syst. 29:207-231.

Forman, R.T.T., and R.D. Deblinger. 2000. The ecological road-effect zone of a Massachusetts (U.S.A.) suburban highway. Conserv. Biol. 14(1):36-46.

Forman, R.T.T., D. Sperling, J.A. Bissonette, A.P. Clevenger, C.C. Cutshal, V.H. Dale, L. Fahrig, R. France, C.R. Coldman, K. Heanue, J.A. Jones, W.J. Swanson, T. Turrentine, and T.C. Winter. 2003. Road Ecology: Science and Solutions. Washington, DC: Island Press.

Fowle, S.C. 1996. The Painted Turtle in the Mission Valley of Western Montana. M.S. Thesis, University of Montana, Missoula, MT.

Franklin, S.E., E.E. Dickson, D.R. Farr, M.J. Hansen, and L.M. Moskal. 2000. Quantification of landscape change from satellite remote sensing. For. Chron. 76(6):877-886.

Fuller, T.K. 1989. Population Dynamics of Wolves in North-Central Minnesota. Wildlife Monograph 105. Bethesda, MD: Wildlife Society.

GAO (U.S. Government Accounting Office). 1994. Highway Planning: Agencies are Attempting to Expedite Environmental Reviews, but Barriers Remain: Report to the Chairman, Subcommittee on Transportation, Committee on Appropriations, House of Representatives. GAO/RCED-94-211. Washington, DC: U.S. Government Accounting Office.

GAO (U.S. Government Accounting Office). 2003. Highway Infrastructure: Stakeholders' Views on the Time to Conduct Environmental Reviews of Highway Projects: Report to the Chairman, Committee on Transportation and Infrastructure, House of Representatives. GAO-03-534. Washington, DC: U.S. Government Accounting Office.

Gelbard, J.L., and J. Belnap. 2003. Roads as conduits for exotic plant invasions in a semiarid landscape. Conserv. Biol. 17(2):420-432.

Gerlach, G., and K. Musolf. 2000. Fragmentation of landscape as a cause for genetic subdivision in bank voles. Conserv. Biol. 14(4):1066-1074.

Gibbs, J.P., and G. Shriver. 2002. Estimating the effects of road mortality on turtle populations. Conserv. Biol. 16(6):1647-1652.

Gjessing, E., E. Lygren, L. Berglind, T. Gulbrandsen, and R. Skanne. 1984. Effect of highway runoff on lake water quality. Sci. Total Environ. 33(1-4):245-257.

Gunderson, L.H., and C.S. Holling, eds. 2002. Panarchy: Understanding Transformations in Human and Natural Systems. Washington, DC: Island Press.

Gunderson, L.H., and J.R. Snyder. 1994. Fire patterns in the southern Everglades. Pp. 291-305 in Everglades: The Ecosystem and Its Restorations, S.M. Davis, and J.C. Ogden, eds. Delray Beach, FL: St. Lucie Press.

Gurney, W., A. Ross, and N. Broekhuizen. 1995. Coupling dynamics of systems and materials. Pp. 176-193 in Linking Species and Ecosystems, C. Jones, and J. Lawton, eds. London: Chapman and Hall.

Gutzwiller, K.J., and W.C. Barrow, Jr. 2003. Bird communities, roads and development: Prospects and constraints of applying empirical models. Biol. Conserv. 113(2):239-243.

Guyot, G., and J. Clobert. 1997. Conservation measures for a population of Hermann's tortoise, Testudo hermanni in southern France bisected by a major highway. Biol. Conserv. 79(2):251-256.

Haas, P.M. 1990. Saving the Mediterranean: The Politics of International Environmental Cooperation. New York: Columbia University Press.

Habron, G.B., M.D. Kaplowitz, and R.L. Levine. 2004. A soft systems approach to watershed management: A road salt case study. Environ. Manage. 33(6):776-787.

Haeuber, R.A., and N.T. Hobbs. 2001. Many small decisions: Incorporating ecological knowledge in land-use decisions in the United States. Pp. 255-

275 in Applying Ecological Principles to Land Management, V.H. Dale, and R.A. Haeuber, eds. New York: Springer.

Hahn, H.H., and R. Pfeifer. 1994. The contribution of parked vehicle emission to the pollution of urban run-off. Sci. Total. Environ. 146/147:525-533.

Hanski, I. 1999. Metapopulation Ecology. Oxford: Oxford University Press.

Harrison, R.M., and S.J. Wilson. 1985. The chemical composition of highway drainage waters. III. Runoff water metal speciation characteristics. Sci. Total Environ. 43(1-2):89-102.

Hawbaker, T.J., and V.C. Radeloff. 2004. Roads and landscape pattern in Northern Wisconsin based on a comparison of four road data sources. Conserv. Biol. 18(5):1233-1244.

Hedrick, P.W. 2004. Genetics of Populations, 3rd Ed. Boston: Jones and Bartlett Publishers.

Heinz (The H. John Heinz III Center for Science, Economics, and the Environment). 2002. The State of the Nation's Ecosystems: Measuring the Lands, Waters, and Living Resources of the United States. New York: Cambridge University Press [online]. Available: http://www.heinzctr.org/ecosystems/report.html [accessed Feb. 11, 2004].

Holling, C.S. 1978. Adaptive Environmental Assessment and Management. Chichester: Wiley.

Holling, C.S. 1986. The Resilience of ecosystems: Local surprise and global change. Pp. 292-320 in Sustainable Development of the Biosphere, W.C. Clark, and R.E. Munn, eds. Cambridge: Cambridge University Press.

Holling, C.S. 1992. Cross-scale morphology, geometry and dynamics of ecosystems. Ecol. Monogr. 62(4):447-502.

Holling, C.S., D.W. Schindler, B.W. Walker, and J. Roughgarden. 1995. Biodiversity in the functioning of ecosystems: An ecological synthesis. Pp. 44-83 in Biodiversity Loss, C. Perrings, K.G. Maler, C. Folke, C.S. Holling, and B.O. Jansson, eds. Cambridge: Cambridge University Press.

Hunsaker, C.T., and D.E. Carpenter, eds. 1990. Environmental Monitoring and Assessment Program: Ecological Indicators. EPA/600/3-90/060. Office of Research and Development, U.S. Environmental Protection Agency, Research Triangle Park, NC.

Ilgen, E., N. Karfich, K. Levsen, J. Angerer, P. Schneider, J. Heinrich, H.E. Wichmann, L. Dunemann, and J. Begerow. 2001. Aromatic hydrocarbons in the atmospheric environment: Part I. Indoor versus outdoor sources, the influence of traffic. Atmos. Environ. 35(7):1235-1252.

Jaarsma, C.F. 1997. Approaches for the planning of rural road networks according to sustainable land use planning. Landscape Urban Plan. 39(1):47-54.

Jaarsma, C.F., and G.P.A. Willems. 2002. Reducing habitat fragmentation by minor rural roads through traffic calming. Landscape Urban Plan. 58(2):125-135.

Jensen, W.F., T.K. Fuller, and W.L. Robinson. 1986. Wolf, Canis lupus, distribution on the Ontario-Michigan border near Sault Ste. Marie. Can. Field Nat. 100(3):363-366.

Johnson, H.B., F.C. Vasek, and T. Yonkers. 1975. Productivity, diversity and stability relationships in Mojave desert roadside vegetation. Bull. Torrey Bot. Club 102(3):106-115.

Johnson, W.C., and S.K. Collinge. 2004. Landscape effects on black-tailed prairie dog colonies. Biol. Conserv. 115(3):487-497.

Jones, J.A., and G.E. Grant. 1996. Peak flow responses to clear-cutting and roads in small and large basins, western Cascades, Oregon. Water Resour. Res. 32(4):959-974.

Jones, J.A., F.J. Swanson, B.C. Wemple, and K.U. Snyder. 2000. Effects of roads on hydrology, geomorphology, and disturbance patches in stream networks. Conserv. Biol. 14(1):76-85.

Jones, M.E. 2000. Road upgrade, road mortality and remedial measures: Impacts on a population of eastern quolls and Tasmanian devils. Wildlife Res. 27(3):289-296.

Karr, J.R. 1981. Assessment of biotic integrity using fish communities. Fisheries 6(6):21-27.

Kasworm, W.F., and T. Manley. 1990. Road and trail influences on grizzly and black bears in northwest Montana. Pp. 79-84 in Bears-Their Biology and Management: 8th International Conference on Bear Research and Management: A selection of papers from the conference held at Victoria, British Columbia, Canada, February 1989, L.M. Darling, and W.R. Archibald, eds. International Conference on Bear Research and Management.

Kautz, R.S., T. Gilbert, and B. Stys. 1999. A GIS plan to protect fish and wildlife resources in the Big Bend area of Florida. Pp. 193-208 in Proceedings of the Third International Conference on Wildlife Ecology and Transportation, G.L. Evink, P. Garrett, and D. Zeigler, eds. FL-ER-73-99. Washington, DC: U.S. Department of Transportation, Federal Highway Administration.

Keller, V.E. 1991. The effect of disturbance from roads on the distribution of feeding sites of geese wintering in north-east Scotland. Ardea 79(2):229-232.

Kelly, J.R., and M.A. Harwell. 1990. Indicators of ecosystem recovery. Environ. Manage. 14(5):527-545.

Kerri, K.D., J.A. Racine, and R.B. Howell. 1985. Forecasting pollutant loads from highway runoff. Transport. Res. Rec. 1017:39-46.

Knaepkens, G., E. Verheyen, P. Galbusera, and M. Eens. 2004. The use of genetic tools for the evaluation of a potential migration barrier for the bullhead. J. Fish Biol. 64(6):1737-1744.

Kremen, C. 1992. Assessing the indicator properties of species assemblages for natural areas monitoring. Ecol. Appl. 2(2):203-217.

Landres, P.B., J. Verner, and J.W. Thomas. 1988. Ecological uses of vertebrate species: A critique. Conserv. Biol. 2(4):316-328.

Larkin, G.A., and K.J. Hall. 1998. Hydrocarbon pollution in the Brunette River watershed. Water Qual. Res. J. Can. 33(1):73-94.

Latha, K.M., and K.V.S. Badarinath. 2003. Black carbon aerosols over tropical urban environment - a case study. Atmos. Res. 69(1-2):125-133.

Levin, S.A. 1999. Towards a science of ecological management. Conserv. Ecol. 3(2):6 [online]. Available: http://www.consecol.org/vol3/iss2/art6 [accessed June 2, 2004].

Light, S.S., L.H. Gunderson, and C.S. Holling. 1995. The Everglades: Evolution of management in a turbulent environment. Pp. 103-168 in Barriers and Bridges to the Renewal of Ecosystems and Institutions, L.H. Gunderson, C.S. Holling, and S.S. Light, eds. New York: Columbia University Press.

Lopes, T. J., and S.G. Dionne. 1998. A Review of Semivolatile and Volatile Organic Compounds in Highway Runoff and Urban Stormwater. Rapid City, SD: U.S. Department of the Interior, U.S. Geological Survey.

Lord, B.N. 1985. Highway runoff/drainage impacts. Pp. 387-390 in Perspectives on Nonpoint Source Pollution: Proceedings of a National Conference, May 19-22, 1985, Kansas City, MO. EPA 440/5-85-001. U.S. Environmental Protection Agency, Washington, DC.

Ludwig, D., R. Hilborn, and C. Walters. 1993. Uncertainty, resource exploitation and conservation: Lessons from history. Science 260(5104):17-36.

MacLean, D.A., P. Etheridge, J. Pelham, and W. Emrich. 1999. Fundy Model Forest: Partners in sustainable forest management. For. Chron. 75(2):219-227.

Maehr, D.S., E.D. Land, and M.E. Roelke. 1991. Mortality patterns of panthers in southwest Florida. Proc. Annu. Conf. SEAFWA 45:201-207 [online]. Available: http://wld.fwc.state.fl.us/critters/panther/index.asp [accessed Feb. 11, 2004].

Mak, K.K., and D.L. Sicking. 2003. Roadside Safety Analysis Program-Engineer's Manuel. NCHRP Report 492. Washington, DC: Transportation Research Board [online]. Available: http://gulliver.trb.org/publications/nchrp/nchrp%5Frpt%5F492.pdf [accessed Jan. 14, 2004].

Manel, S., M.K. Schwartz, G. Luikart, and P. Taberlet. 2003. Landscape genetics: Combining landscape ecology and population genetics. Trends Ecol. Evol. 18(4):189-197.

Maurer, M. 1999. Development of a community-based, landscape-level terrestrial mitigation decision support system for transportation planners. Pp. 99-110 in Proceedings of the Third International Conference on Wildlife Ecology and Transportation, G.L. Evink, P. Garrett, and D. Zeigler, eds. FL-ER-73-99. Washington, DC: U.S. Department of Transportation, Federal Highway Administration.

Mech, L.D., S.H. Fritts, G.L. Radde, and W.J. Paule. 1988. Wolf distribution and road density in Minnesota. Wildl. Soc. Bull. 16:85-87.

Merrill, S.B. 2000. Road densities and gray wolf, Canis lupus, habitat suitability: An exception. Can. Field Nat. 114(2):312-313.

Millennium Ecosystem Assessment. 2003. Ecosystems and Human Well-being: A Framework for Assessment. Washington, DC: Island Press.

Mills, L.S., M.E. Soulé, and D.F. Doak. 1993. The key-stone species concept in ecology and conservation. BioScience 43(4):219-224.

Ministry of Transport, P.W.A.W.M. 1994. The Chemical Quality of Verge Grass in the Netherlands. No. 62. Dienst Wegen Waterbouwkunde, Ministerie van Verkaar en Waterstaat, Delft, Netherlands.

Mladenoff, D.J., T.A. Sickley, R.G. Haight, and A.P. Wydeven. 1995. A regional landscape analysis and prediction of favorable wolf habitat in the Northern Great Lakes region. Conserv. Biol. 9(2):279-294.

Mladenoff, D.J., T.A. Sickley, and A.P. Wydeven. 1999. Predicting gray wolf landscape recolonization: Logistic regression models vs. new field data. Ecol. Appl. 9(1):37-44.

Moore, K., M. Furniss, S. Firor, and M. Love. 1999. Fish Passage through Culverts: An Annotated Bibliography. Six Rivers National Forest Watershed Interactions Team, Eureka, CA. November 5, 1999 [online]. Available: http://stream.fs.fed.us/fishxing/biblio.html [accessed April 15, 2005].

Morrison, P., P. Morrison, and The Office of Charles and Ray Eames. 1982. Power of Ten. New York: Scientific American Library.

MSHA (Maryland State Highway Administration). 1998. Thinking Beyond the Pavement: A National Workshop on Integrating Highway Development with Communities and the Environment While Maintaining Safety and Performance. Maryland Department of Transportation State Highway Administration [online]. Available: http://www.fhwa.dot.gov/csd/mdbroch.pdf [accessed March 4, 2005].

National Invasive Species Council. 2004. What Is an Invasive Species? National Agricultural Laboratory [online]. Available: http://www.invasivespecies.gov/ [accessed May 26, 2004].

Nellemann, C., and R.D. Cameron. 1998. Cumulative impacts of an evolving oil-field complex on the distribution of calving caribou. Can. J. Zool. 76(8):1425-1430.

Nellemann, C., I. Vistnes, P. Jordhøy, and O. Strand. 2001. Winter distribution of wild reindeer in relation to power lines, roads and resorts. Biol. Conserv. 101(3):351-360.

Nelson, P.O., W.C. Huber, N.N. Eldin, K.J. Williamson, J.R. Lundy, M.F. Azizian, P. Thayumanavan, M.M. Quigley, E.T. Hesse, K.M. Frey, and R.B. Leahy. 2001. Environmental Impact of Construction and Repair Materials on Surface and Ground Waters: Summary of Methodology, Laboratory Results, and Model Development. National Cooperative

Highway Research Program Report 448. Washington, DC: National Academy Press.

NHDOT (New Hampshire Department of Transportation). 2004. Traffic congestion. Pp. 21-23 in Ten Year Transportation Improvement Plan 2005-2014 [online]. Available: http://www.state.nh.us/dot/transportation planning/pdf/04Congest.pdf [accessed March 2, 2005].

Noon, B.R., T.A. Spies, and M.G. Raphael. 1999. Conceptual basis for designing an effectiveness monitoring program. Pp. 21-48 in The Strategy and Design of the Effectiveness Monitoring Program for the Northwest Forest Plan, B.S. Mulder, B.R. Noon, T.A. Spies, M.G. Raphael, C.J. Palmer, A.R. Olsen, G.H. Reeves, and H.H. Welsh, eds. Gen. Tech. Rep. PNW-GTR-437. Portland, OR: U.S. Department of Agriculture, Forest Service, Pacific Northwest Research Station [online]. Available: http://www.fs.fed.us/pnw/pubs/gtr%5F437.pdf [accessed Feb. 25, 2004].

Noss, R.F. 1990. Indicators for monitoring biodiversity: A hierarchical approach. Conserv. Biol. 4(4):355-364.

Noss, R.F., and A. Cooperrider. 1994. Saving Nature's Legacy: Protecting and Restoring Biodiversity. Washington, DC: Island Press.

Noss, R.F., H.B. Quigley, M.G. Hornocker, T. Merrill, and P.C. Paquet. 1996. Conservation biology and carnivore conservation in the Rocky Mountains. Conserv. Biol. 10(4):949-963.

NRC (National Research Council). 1995. Science and the Endangered Species Act. Washington, DC: National Academy Press.

NRC (National Research Council). 1996. River Resource Management in the Grand Canyon. Washington, DC: National Academy Press.

NRC (National Research Council). 1999. Downstream: Adaptive Management of Glen Canyon Dam and the Colorado River Ecosystem. Washington, DC: National Academy Press.

NRC (National Research Council). 2000a. Global Change Ecosystem Research. Washington, DC: National Academy Press.

NRC (National Research Council). 2000b. Ecological Indicators for the Nation. Washington, DC: National Academy Press.

NRC (National Research Council). 2001. Compensating for Wetlands Losses Under the Clean Water Act. Washington, DC: National Academy Press.

NRC (National Research Council). 2002. Predicting Invasions of Nonindigenous Plants and Plant Pests. Washington DC: National Academy Press.

NRC (National Research Council). 2003. Cumulative Environmental Effects of Oil and Gas Activities on Alaska's North Slope. Washington, DC: National Academies Press.

NRC (National Research Council). 2004. Atlantic Salmon in Maine. Washington, DC: National Academies Press.

Odum, H.T. 1983. Systems Ecology: An Introduction. New York: Wiley.

Odum, H.T. 1996. Environmental Accounting: EMERGY and Environmental Decision Making. New York: Wiley.

O'Neill, R.V., D.L. DeAngelis, J.B.Waide, and T.F.H. Allen. 1986. A Hierarchical Concept of Ecosystems. Monographs in Population Biology 23. Princeton, NJ: Princeton University Press.

Ortega, Y.K., and D.E. Capen. 1999. Effects of forest roads on habitat quality for ovenbirds in a forested landscape. Auk 116(4):937-946.

Ostrom, L. 1990. Governing the Commons: The Evolution of Institutions for Collective Action. New York: Cambridge University Press.

Parendes, L.A., and J.A. Jones. 2000. Role of light availability and dispersal in exotic plant invasion along roads and streams in the H.J. Andrews Experimental Forest, Oregon. Conserv. Biol. 14(1):64-75.

Pedlowski, M.A., V.H. Dale, E.A.T. Matricardi, and E.P. da Silva Filho. 1997. Patterns and impacts of deforestation in Rondonia, Brazil. Landscape Urban Plan. 38(3-4):149-157.

Peterson, G.D. 2002. Forest dynamics in the southeastern United States: Managing multiple stable states. Pp. 227-246 in Resilience and the Behavior of Large Scale Systems, L.H. Gunderson, and L. Pritchard, Jr., eds. Washington DC: Island Press.

Piepers, A.A.G., ed. 2001. Infrastructure and Nature; Fragmentation and Defragmentation. Dutch State of the Art Report for COST Activity 341. Defragmentation Series, Part 39A [in Dutch]. Road and Hydraulic Engineering Division, Delft, the Netherlands.

Price, S. 2003. Transport and the Environment- The Implications of the 10 Year Plan. Presentation at Road and Mammals, the Annual Autumn Symposium of the Mammal Society held jointly with the Highways Agency, Nov. 28, 2003, London, UK.

Raskin, P., G. Gallopin, P. Gutman, A. Hammond, and R. Stewart. 1998. Bending the Curve: Toward Global Sustainability. Polestar Series Report No. 8. Stockholm Environment Institute [online]. Available: http://www.tellus.org/seib/publications/bendingthecurve.pdf [accessed August 23, 2005].

Reijnen, R., and R. Foppen. 1994. The effects of car traffic on breeding bird populations in woodland. I. Evidence of reduced habitat quality for willow warblers (Phylloscopus trochilus) breeding close to a highway. J. Appl. Ecol. 31(1):85-94.

Reijnen, R., and R. Foppen. 1995. The effect of car traffic on breeding bird population in woodland: 4. Influence of population size on the reduction of density close to a highway. J. Appl. Ecol. 32(3):481-491.

Reijnen, R., R. Foppen, C. ter Braak, and J. Thissen. 1995. The effects of car traffic on breeding bird populations in woodland: III. Reduction of density in relation to the proximity of main roads. J. Appl. Ecol. 32(1):187-202.

Reijnen, R., R. Foppen, and H. Meeuwsen. 1996. The effects of car traffic on the density of breeding birds in Dutch agricultural grasslands. Biol. Conserv. 75(3):255-260.

Reijnen, R., R. Foppen, and G. Veenbaas. 1997. Disturbance by traffic of breeding birds: Evaluation of the effect and considerations in planning and managing road corridors. Biodivers. Conserv. 6(4):567-581.

Richburg, J.A., W.A. Patterson III, and F. Lowenstein. 2001. Effects of road salt and Phragmites australis invasion of the vegetation of a western Massachusetts calcerous lake-basin fen. Wetlands 21(2):247-255.

Robison, E.G., A.M. Mirati, and M. Allen. 1999. Oregon Road/Stream Crossing Restoration Guide. Oregon Department of Fish and Wildlife. June 8, 1999 [online]. Available: http://www.dfw.state.or.us/odfwhtml/ infocntrfish/management/oregonrd_stream.pdf [accessed April 15, 2005].

Romin, L.A., and J.A. Bissonette. 1996. Deer-vehicle collisions: Status of state monitoring activities and mitigation efforts. Wildlife Soc. B. 24(2): 276-283.

Rost, G.R., and J.A. Bailey. 1979. Distribution of mule deer and elk in relation to roads. J. Wildl. Manage. 43(3):634-641.

Saitoh, T.S., T. Shimada, and H. Hoshi. 1996. Modeling and simulation of the Tokyo urban heat island. Atmos. Environ. 30(20):3431-3442.

Sansalone, J.J., K.P. Paris, and S.G. Buchberger. 1995. Heavy metals in the first-flush of highway nonpoint pollutants. Pp. 1375-1379 in Water Resources Engineering: Proceedings of the First International Conference, August 14-18, 1995, San Antonio, W.H. Espey, and P.G. Combs, eds. New York: American Society of Civil Engineers.

Schaefer, J.F., E. Marsh-Matthews, D.E. Spooner, K.B. Gido, and W.J. Matthews. 2003. Effects of barriers and thermal refugia on local movement of the threatened leopard darter, Percina pantherina. Environ. Biol. Fish. 66(4):391-400.

Scheffer, M. 1998. Ecology of Shallow Lakes. London: Chapman and Hall.

Schlesinger, W.H. and C.S. Jones. 1984. The comparative importance of overland runoff and mean annual rainfall to shrub communities of the Mojave Desert. Bot. Gaz. 145:116-124.

Schrank, D., and T. Lomax. 2001. 2001 Urban Mobility Study. College Station, TX: Texas Transportation Institute. May 2001.

Scigliano, E. 2003. Turn down the lights. Discover 24(7):60-63.

Servheen, C., J. Waller, and W. Kasworm. 1998. Fragmentation effects of high-speed highways on grizzly bear populations shared between the United States and Canada. Pp. 97-103 in Proceedings of the International Conference on Wildlife Ecology and Transportation (ICOWET), February 10-12, 1998, Fort Meyers, FL, G.L. Evink, P. Garrett, D. Zeigler, and J. Berry, eds. FL-ER-69-98. Washington, DC: U.S. Department of Transportation, Federal Highway Administration.

Shabecoff, P. 1989. EPA to ban virtually all asbestos products by '96. New York Times, Sec. A, P. 8. July 7, 1989.

Shaheen, D.G. 1975. Contributions of Urban Roadway Usage to Water Pollution. EPA-600/2-75-004. Office of Research and Development, U.S. Environmental Protection Agency, Washington, DC.

Shallat, T.A. 1994. Structures in the Stream: Water, Science and the Rise of the U.S. Army Corps of Engineers, 1st Ed. Austin, TX: University of Texas Press.

Sharpe, G.W. 2004. Intersection Design Guidelines. Memorandum to Chief District Engineers, Design Engineers, and Active Consultants, from Gary W. Sharpe, Director, Division of Highway Design, Kentucky Transportation Cabinet, Frankfort, KY. Design Memorandum No. 2-04. February 4, 2004 [online]. Available: http://www.kytc.state.ky.us/design/memos/DesignMemo2_042.pdf [accessed August 31, 2004].

Shepherd, J.M., H. Pierce, and A.J. Negri. 2002. Rainfall modification by major urban areas: Observations from spaceborne rain radar on the TRMM satellite. J. Appl. Meteorol. 41(7):689-701.

Sheppe, W. 1965. Island populations and gene flow in the deer mouse, Peromyscus leucopus. Evolution 19(4):480-495.

Sherwin, R.E., W.L. Gannon, and S. Haymond. 2000. The efficacy of acoustic techniques to infer differential use of habitat by bats. Acta Chiropterol. 2(2):145-153.

Shrouds, J. 2001. Migratory Bird Treaty Act and Executive Order 13186. Letter to Division Administrators, Federal Lands Highway Division Engineers, Directors of Field Services, from James Shrouds, Director, Office of Natural Environment, Federal Highway Administration, U.S. Department of Transportation. February 2, 2001 [online]. Available: http://www.fhwa.dot.gov/environment/migbird.htm [accessed June 1, 2004].

Simberloff, D. 1997. Flagships, umbrellas, and keystones: Is single-species management passé in the landscape era? Biol. Conserv. 83(3):247-257.

Simon, H.A. 1962. The architecture of complexity. Proc. Am. Philos. Soc. 106(6):467-482.

Slatkin, M. 1987. Gene flow and the geographic structure of natural populations. Science 236(4803):787-792.

Slocombe, D.S. 1998. Defining goals and criteria for ecosystem-based management. Environ. Manage. 22(4):483-493.

Smith, D.J. 1999. Identification and prioritization of ecological interface zones on state highways in Florida. Pp. 209-230 in Proceedings of the Third International Conference on Wildlife Ecology and Transportation, G.L. Evink, P. Garrett, and D. Zeigler, eds. FL-ER-73-99. Washington, DC: U.S. Department of Transportation, Federal Highway Administration.

Stohlgren, T.J., G.W. Chong, M.A. Kalkhan, L.D. Schell. 1997. Rapid assessment of plant diversity patterns: A methodology for landscapes. Environ. Monit. Assess. 48(1):25-43.

Stork, N.E., T.J.B. Boyle, V. Dale, H. Eeley, B. Finegan, M. Lawes, N. Manokaran, R. Prabhu, and J. Soberon. 1997. Criteria and Indicators for Assessing Sustainability of Forest Management: Conservation of Biodiversity. Center for International Forestry Research Working Paper No. 17. Jakarta, Indonesia: Center for International Forestry Research [online]. Available: http://www.cifor.cgiar.org/publications/pdf_files/WPapers/WP-17.pdf [accessed July 22, 2005].

STPP (Surface Transportation Policy Project). 2002. Transportation Project Delays: Why Environmental "Streamlininig" Won't Solve the Problem. Decoding Transportation Policy and Practice No. 6 [online]. Available: http://www.stpp.org/library/decoding.asp [accessed March 15, 2005]

Suter, G. 1993. Ecological Risk Assessment. Boca Raton, FL: Lewis.

Sutter, P.S. 2002. Driven Wild: How the Fight against Automobiles Launched the Modern Wilderness Movement. Seattle: University of Washington Press.

Suttle, K.B., M.E. Power, J.M. Levine, and C. McNeely. 2004. How fine sediment in riverbeds impairs growth and survival of juvenile salmonids. Ecol. Appl. 14(4):969-974.

Swihart, R.K., and N.A. Slade. 1984. Road crossing in Sigmodon hispidus and Microtus ochrogaster. J. Mammal. 65(2):357-360.

Taulman, J.F., and L.W. Robbins. 1996. Recent range expansion and distributional limits of the 9-banded armadillo (Dasypus novemcinctus) in the United-States. J. Biogeogr. 23(5):635-648.

Thiel, R.P. 1985. Relationship between road densities and wolf habitat suitability in Wisconsin. Am. Midl. Nat. 113(2):404-407.

Thompson, L.M. 2003. Abundance and Genetic Structure of Two Black Bear Populations Prior to Highway Construction in Eastern North Carolina. M.S. Thesis, University of Tennessee, Knoxville.

Tinker, D.B., C.A.C. Resor, G.P. Beauvais, K.F. Kipfmueller, C.I. Fernandes, and W.L. Baker. 1998. Watershed analysis of forest fragmentation by clearcuts and roads in a Wyoming forest. Landscape Ecol. 13(3):149-165.

TPF (Transportation Pooled Fund). 2004. Transportation Pooled Fund Program [online]. Available: http://www.pooledfund.org/ [accessed May 26, 2004].

TranSafety, Inc. 1997. Designing highway culverts that do not impede the movements of resident fish species. Road Engineering Journal, November 1, 1997 [online]. Available: http://www.usroads.com/journals/p/rej/9711/re971102.htm [accessed April 14, 2005].

TransTech Management, Inc. 2000. Environmental Streamlining: A Report on Delays Associated with the Categorical Exclusion and Environmental Assessment Processes. Prepared for American Association of State Highway and Transportation Officials Standing Committee on Highways and the National Cooperative Highway Research Program, by TransTech Management, Inc., Washington, DC. October 2000.

TRB (Transportation Research Board). 2002a. Surface Transportation Environmental Research: A Long Term Strategy. Special report 268. Washington, DC: National Academy Press.

TRB (Transportation Research Board). 2002b. Environmental Research Needs in Transportation. Conference Proceedings 28. Washington, DC: National Academy Press.

TRB (Transportation Research Board). 2003. Roadside Safety Analysis Program-Engineer's Manual. RSAP Software and User's Manual. TRB National Cooperative Highway Research Program 22-9 [online]. Available: http://trb.org/news/blurb_detail.asp?id=1519 [accessed Jan. 14, 2005].

TRB (Transportation Research Board). 2004. Evaluation of the Use and Effectiveness of Wildlife Crossings. National Cooperative Highway Research Program Project 25-27 [online]. Available: http://www4.trb.org/trb/crp.nsf/0/29f4dcb2345e34d685256d0b0065f79d?OpenDocument [accessed August 18, 2004].

Treweek, J.R., P. Hankard, D.B. Roy, H. Arnold, and S. Thompson. 1998. Scope for strategic ecological assessment of trunk-road development in England with respect to potential impacts on lowland heathland, the Dartford warbler (Sylvia undata) and the sand lizard (Lacerta agilis). J. Environ. Manage. 53(2):147-163.

Trombulak, S.C., and C.A. Frissell. 2000. Review of ecological effects of road on terrestrial and aquatic communities. Conserv. Biol. 14(1):18-30.

Tsai, P.J., C.C. Lee, M.R. Chen, T.S. Shih, C.H. Lai, and S.H. Liou. 2002. Predicting the contents of BTEX and MTBE for the three types of tollbooth at a highway toll station via the direct and indirect approaches. Atmos. Environ. 36(39-40):5961-5969.

Turnpenny, A. and J. Nedwell, J. Maes, C. Taylor, and D. Lambert. 2003. Bubble Curtain Technology. Development and Operation of Acoustic Fish Deterrent Systems at Estuarine Power Stations. Presentation at EPA Symposium on Cooling Water Intake Technologies to Protect Aquatic Organism, May 6-7, 2003, Arlington, VA [online]. Available: http://www.epa.gov/waterscience/presentations/session_d-2/turnpenny.pdf

Turtle, S.L. 2000. Embryonic survivorship of the spotted salamander (Ambystoma maculatum) in roadside and woodland vernal pools in southeastern New Hampshire. J. Herpetol. 34(1):60-67.

Tyser, R.W., and C.A. Worley. 1992. Alien flora in grasslands adjacent to road and trail corridors in Glacier National Park, Montana (USA). Conserv. Biol. 6(2):253-262.

University of Florida. 2001. The EPA Southeastern U.S. Ecological Framework Project. GeoPlan Center, University of Florida, Region 4 Planning & Analysis Branch, U.S. Environmental Protection Agency [online]. Available: http://www.geoplan.ufl.edu/epa/ [accessed Feb. 20, 2004].

USACE (U.S. Army Corps of Engineers). 2002. Regulatory Guidance Letter. Guidance on Compensatory Mitigation Projects for Aquatic Resource Impacts Under the Corps Regulatory Pursuant to Section 404 of the Clean Water Act and Section 10 of the Rivers and Harbors Act of 1899. No. 02-

2. December 24, 2002 [online]. Available: http://www.epa.gov/owow/ wetlands/pdf/RGL_02-2.pdf [accessed Feb. 17, 2004].

USFWS (U.S. Fish and Wildlife Service). 1980. Habitat Evaluation Procedures (HEP). Ecological Service Manual ESM 103. Fort Collins, CO: U.S. Fish and Wildlife Service.

USGS (U.S. Geological Survey). 2004. Seamless Data Distribution [online]. Available: http://gisdata.usgs.net/website/seamless/viewer.php [accessed June 6, 2005].

van Bohemen, H., C. Padmos, and H. de Vries. 1994. Versnippering-ontsnippering: Beleid en onderzoek bij verkeer en waterstaat. Landschap 1994(3):15-25.

van der Heijden, K. 1996. 1965 to 1990: Five discoveries at Shell. Chapter 1 in Scenario: The Art of Strategic Conversation, 1st Ed. New York: John Wiley & Sons [online]. Available: http://media.wiley.com/product_data/ excerpt/86/04700236/0470023686.pdf [accessed August 23, 2005].

van der Zande, A.N., W.J. ter Kers, and W.J. van der Weijden. 1980. The impact of roads on the densities of four bird species in an open field habitat—evidence of a long-distance effect. Biol. Conserv. 18(4):299-321.

van Dyke, F.G., R.H. Brocke, and H.G. Shaw. 1986. Use of road track counts as indices of mountain lion presence. J. Wildl. Manage. 50(1):102-109.

Viskari, E.L., M. Vartiainen, and P. Pasanen. 2000. Seasonal and diurnal variation in formaldehyde and acetaldehyde concentrations along a highway in Eastern Finland. Atmos. Environ. 34(6):917-923.

Vos, C.C., and J.P. Chardon. 1998. Effects of habitat fragmentation and road density on the distribution pattern of the moor frog Rana arvalis. J. Appl. Ecol. 35(1):44-56.

Walker, B.H. 2002. Ecological resilience in grazed rangelands: A generic case study. Pp. 183- 193 in Resilience and the Behavior of Large Scale Systems, L.H. Gunderson, and L. Pritchard Jr., eds. Washington, DC: Island Press.

Walker, D.A., P.J. Webber, E.F. Binnian, K.R. Everett, N.D. Lederer, E.A. Nordstrand, and M.D. Walker. 1987. Cumulative impacts of oil fields on northern Alaskan landscapes. Science 238:757-761.

Walters, C.J. 1986. Adaptive Management of Renewable Resources. New York: McMillan.

Walters, C. 1997. Challenges in adaptive management of riparian and coastal ecosystems. Conserv. Ecol. 1(2):1 [online]. Available: http://www. consecol.org/vol1/iss2/art1 [accessed June 7, 2004].

Walters, C.J., and C.S. Holling. 1990. Large-scale management experiments and learning by doing. Ecology 71(6):2060-2068.

Ward, A.L. 1982. Mule deer behavior in relation to fencing and underpasses on Interstate 80 in Wyoming. Transport. Res. Rec. 859:8-13.

Warren, M.L., Jr., and M.G. Pardew. 1998. Road crossings as barriers to small-stream fish movement. Trans. Am. Fish. Soc. 127(4):637-644.

Wear, D.N., and P. Bolstad. 1998. Land-use changes in southern Appalachian landscapes: Spatial analysis and forecast evaluation. Ecosystems 1(6):575-594.

Wegner, W., and M. Yaggi. 2001. Environmental impacts of road salt and alternatives in the New York City Watershed. Stormwater 2(5):24-31 [online]. Available: http://www.forester.net/sw_0107_environmental.html [accessed March 11, 2005].

Werner, W.E. 1956. Mammals of the thousand island region, New York. J. Mammal. 37(3):395-406.

Westley, F. 2002. The devil in the dynamics: Adaptive management on the front lines. Pp. 333-360 in Panarchy: Understanding Transformations in Human and Natural Systems, L.H. Gunderson, and C.S. Holling, eds. Washington, DC: Island Press.

White, P.A., and M. Ernst. 2003. Second Nature: Improving Transportation Without Putting Nature Second. Surface Transportation Policy Project. Defenders of Wildlife [online]. Available: http://www.transact.org/library/reports_pdfs/Biodiversity/second_nature.pdf [accessed Feb. 27, 2004].

Wilhelm, M., and B. Ritz. 2003. Residential proximity to traffic and adverse birth outcomes in Los Angeles County, California, 1994-1996. Environ. Health Perspect. 111(2):207-216.

Wilkie, D., E. Shaw, F. Rotberg, G. Morelli, and P. Auzel. 2000. Roads, development, and conservation in the Congo basin. Conserv. Biol. 14(6):1614-1622.

Woolum, C.A. 1964. Notes from a Study of the Microclimatology of the Washington, DC Area for the Winter and Spring Seasons. Weatherwise 17(6).

Wright, S.H. 1989. Sourcebook of Contemporary North American Architecture: From Postwar to Postmodern. New York: Van Nostrand Reinhold.

WSDOT (Washington State Department of Transportation). 2003. WSDOT Traffic and Weather. Washington State Department of Transportation [online]. Available: http://www.wsdot.wa.gov/traffic [accessed Feb. 13, 2004].

Wu, J.S., C.J. Allan, W.L. Saunders, and J.B. Evett. 1998. Characterization and pollutant loading estimation for highway runoff. J. Environ. Eng. 124(7):584-592.

Yanes, M., J.M. Velasco, and F. Suárez. 1995. Permeability of roads and railways to vertebrates: The importance of culverts. Biol. Conserv. 71(3):217-222.

Yousef, Y.A., H.H. Harper, L.P. Wiseman, and J.M. Bateman. 1985. Consequential species of heavy metals in highway runoff. Transport. Res. Rec. 1017:56-62.

Zmirou, D., S. Gauvin, I. Pin, I. Momas, F. Sahraoui, J. Just, Y. Le Moullec, F. Bremont, S. Cassadou, P. Reungoat, M. Albertini, N. Lauvergne, M. Chiron, and A. Labbe. 2004. Traffic related air pollution and incidence of childhood asthma: Results of the Vesta case-control study. J. Epidemiol. Community Health 58(1):18-23.

# Appendix A

# Biographical Information on Committee Members

**Lance Gunderson** (*Chair*) is an associate professor at Emory University. He earned a B.S. and M.S. in botany and a Ph.D. in environmental engineering sciences from the University of Florida. His research interests include how ecosystem processes and structures interact across space and time and how scientific understanding influences resource policy and management. Dr. Gunderson worked for over a decade as a botanist with the U.S. National Park Service in the Big Cypress and Everglades regions of southern Florida. He then worked for a decade as a research scientist in the Department of Zoology at the University of Florida. He has been chairman of the Department of Environmental Studies at Emory University since January 1999. From 1997 to 2000, he served as the executive director of the Resilience Network, a program of the Beijer International Institute for Ecological Economics, Swedish Royal Academy of Sciences, Stockholm. He is currently vice-chair of the Resilience Alliance, and co-editor in chief of Conservation Ecology.

**Anthony Clevenger** is a research scientist at the Western Transportation Institute at Montana State University. He earned a B.S. in the conservation of natural resources from the University of California, Berkeley (1980), M.S. in wildlife ecology from the University of Tennessee (1986), and Ph.D. in vertebrate zoology from the Universidad de León, Spain (1990). He has been studying road effects on wildlife populations in Banff and the surrounding national and provincial parks in the Canadian Rocky Mountains since 1996. Dr. Clevenger's research is focused on long-term monitoring of wildlife crossings to assess their conservation value on animal populations and their performance at maintaining regional landscape connectivity. He has worked as a research wildlife biologist for the World Wide Fund for Nature–International (Gland, Switzerland), Ministry of Environment–France (Toulouse), U.S. Forest Service, and U.S. National Park Service. Dr. Clevenger is the co-author of three books and has published more than 40 scientific articles.

**Adrienne Cooper** is an assistant professor of civil and environmental engineering at Temple University where her research is focused on photocatalytic processes for environmental systems, environmentally sustainable engineering and development, landfill bioreactors, and enzymatic synthesis of chemical compounds. Dr. Cooper earned a B.S. in chemical engineering from the University of Tennessee (1984) and a Ph.D. in environmental engineering from the University of Florida (1998). She also participates in the Sloan Minority Ph.D. Program, which seeks to increase the number of underrepresented minority Ph.D. graduates in civil and environmental engineering through personal mentoring and fellowships. Previously, she was an assistant professor of civil and environmental engineering at the University of South Carolina (1998-2003) and worked for the Alachua County Environmental Protection Department, Gainesville, FL (1994-1995) and for E. I. DuPont de Nemours and Company, Inc., Wilmington, DE (1984-1992).

**Virginia Dale** is corporate fellow at Oak Ridge National Laboratory. She earned her B.A. in mathematics from the University of Tennessee (1974), M.S. in mathematics with a minor in ecology from the University of Tennessee, Knoxville (1975), and Ph.D. in mathematical ecology from the University of Washington (1980). Dr. Dale's primary research interests include environmental decision making, forest succession, land-use change, landscape ecology, and ecological modeling. She has worked on developing tools for integrating socioeconomic and ecological models of land-use change. She is editor-in-chief of the journal *Environmental Management*. She has published over 140 scientific articles and edited six books. She served on the National Academy of Sciences Ecosystem Panel, which produced the report *Global Change Ecosystems Research* and on the editorial boards of three journals. Dr. Dale was also a member of the "Committee of Scientists" appointed by Secretary of Agriculture and the Ecosystems Panel of the National Science Foundation.

**Leonard Evans** is president of the Science Serving Society. He earned his D.Phil. in physics from Oxford University. His primary research interest is traffic safety. He is author of "Traffic Safety" (2004) and "Traffic Safety and the Driver" (1991), and 150 journal articles. He has presented the keynote address at a dozen international traffic safety meetings and has received awards from the International Traffic Medicine Association, the Association for the Advancement of Automotive Medicine, the Human Factors and Ergonomics Society, and National Highway Traffic Safety for his contributions to traffic safety research. He is a member of the National Academy of Engineering.

**Gary Evink** (retired) earned a B.S. in wildlife ecology (1972) and M.S. in aquatic ecology from the University of Florida (1976). He worked for over 31 years as the state ecologist for the Florida Department of Transportation. Mr. Evink managed environmental services in the Environmental Management Office and was involved in all aspects of policy, procedure, training, research, and quality assurance. He has published hundreds of papers on diverse ecological topics and has experience in every area of environmental documentation in the state and federal processes. Mr. Evink developed and carried out the Florida Department of Transportation's Ecosystem Management Program, which was a multidisciplinary approach to a systems-level consideration of environmental issues related to transportation. He chaired the annual International Conference on Wildlife Ecology and Transportation. Mr. Evink participated in five National Cooperative Highway Research Program (NCHRP) panels dealing with wetlands, stormwater, and other ecological issues and chaired two of these panels. He is chairing NCHRP 25-09, which is dealing with the impacts of stormwater from construction and repair materials. Mr. Evink is also a consultant for NCHRP Project 20-5, Synthesis Topic 32-11: "Interactions between wildlife ecology and roadways." He served on the Governor's Commission for a Sustainable South Florida, the South Florida Ecosystem Restoration Task Force, and the State Committee for Environmental Education (chair). He received the 2001 FHWA Environmental Excellence Award for individual leadership.

**Lenore Fahrig** is a professor of biology at Carleton University. She earned a B.S. from Queen's College, M.S. from Carleton University, and Ph.D. from the University of Toronto. Her research is focused on the effects of landscape structure on abundance, distribution, and persistence of organisms. Dr. Fahrig uses spatial simulation modeling to formulate predictions and test these predictions using a range of different organisms. She has been a member of the Carleton University faculty since July 1991.

**Kingsley Haynes** is university professor and dean of the School of Public Policy at George Mason University. He earned a B.A. in history, geography, and political science from Western Michigan University, M.A. in geography from Rutgers University, and Ph.D. in geography and environmental engineering from the Johns Hopkins University. His areas of expertise include resource and environmental management, urban and regional economic development, economic geography, regional science, social systems modeling, policy analysis and governmental affairs. He has served as professor and director of the Regional Economic Development Institute, School of Public and Environmental Affairs at Indiana University, professor of public affairs and geography at the LBJ School of the University of Texas at Austin, and chair of the Department of Geography at Boston University. Dr. Haynes has been involved in regional economic development, natural resource management, transportation and environmental planning in Montana's Yellowstone Basin, in the Lake Michigan and Ohio River regions of Indiana, in the Texas coastal zone and internationally in Brazil, Malaysia, Taiwan, and the Middle East. Using mathematical programming techniques for evaluating resource utilization and economic simulation for community water supply alternatives, he has been active in state resource assessment in Texas, Indiana, and Massachusetts. He also directed the Ford Foundation's Office of Resources and Environment.

**Wayne W. Kober** is president of Wayne W. Kober, Inc. He earned a B.S. in environmental resource management from Pennsylvania State University (1973). He has over 30 years of multimodal transportation and environmental management experience for the public and private sectors. He is nationally recognized as an innovative leader in the field of transportation project development and environmental management. His broad span of experience at successfully integrating environmental analysis, agency and public involvement, and context-sensitive design aspects into a systematic decision-making process has enabled him to play a prominent role in advancing multimodal transportation improvement programs across the country. He led the development of the environmental streamlining and stewardship legislation, policies and practices at the mid-Atlantic state and national levels. From an ecological perspective, he was instrumental in the development and deployment of the Pennsylvania modified habitat evaluation procedure (PAMHEP) used to assess highway wildlife habitat impacts for over an 18-year period on every major transportation project in Pennsylvania, as well as the development and implementation of the Pennsylvania Department of Transportation's wetland and upland mitigation policies and programs. He contributed to the book *Road Ecology—Science and Solutions*. He is in private practice and currently serves as a senior transportation and environmental management technical expert for the American Association of State Highway and Transportation Officials.

**Stephen Lester** is vice president of Urban Engineers, Inc. He earned a Bachelors of Civil Engineering (1965) and Masters of Civil Engineering (1969) from Villanova University. Mr. Lester brought 30 years of experience to the position of vice president of engineering and quality services for Urban Engineers, Inc. when he retired from the position of district engineer for the Pennsylvania Department of Transportation (Penn-DOT). As district engineer, he oversaw the highway and bridge designs, construction and maintenance programs for PennDOT's most populous district. He supervised over 1,000 employees within a five-county area, which included more than 3,900 miles of state highways and 2,600 state-owned bridges. Mr. Lester had previously served as assistant district engineer for maintenance, district traffic engineer, and assistant district locations engineer.

**Kent Redford** is vice president of the Wildlife Conservation Society. He earned a Ph.D. in biology from Harvard University (1987). His research interests range from mammalian biology to traditional peoples and biodiversity conservation. His earlier work concentrated on South American mammals with emphasis on anteaters and armadillos and zoogeographic patterns. He is also interested in the study of subsistence hunting and the impacts it has on the conservation of game species in particular and on biodiversity in general. Dr. Redford's work has been expanded to include the role of parks in protecting biodiversity and the relationships between traditional peoples, resource harvesting, and biodiversity conservation.

**Margaret Strand** is an attorney with Venable, LLP. She earned a B.A. in history from the University of Rochester, M.A. in history from the University of Rhode Island, and J.D. from Marshall Wythe School of Law at the College of William and Mary. Her areas of expertise include environmental policy and wetlands regulation and enforcement. She advises clients on environmental compliance and conducts environmental litigation, focusing on EPA-administered regulatory programs. Before entering private practice, Ms. Strand was chief of the Environmental Defense Section of the U.S. Department of Justice, where she was involved in environmental policy issues, including wetlands regulation and enforcement. Ms. Strand is chair of the Wetlands and Water Quality Committee of the American Bar Association Section of Environment, Energy and Resources Law.

**Paul Wagner** is biology program manager at the Washington Department of Transportation. He earned a B.S. from Juniata College (1986). He has served on a number of review panels, including a National Cooperative Highway Research Program synthesis study of ecology and transportation issues. Mr. Wagner participated in a technical oversight committee that developed a set of ecological models and methodologies for wetland functions assessment for Washington based on the hydrogeomorphic method concepts.

**J.M. Yowell** (retired) was state highway engineer at the Kentucky Transportation Cabinet. He earned a B.S. in civil engineering from the University of Kentucky (1959). Mr. Yowell has 45 years of experience in the transportation industry and has held such positions as project manager for an engineering and construction firm and vice-president for multimillion dollar construction firms. He has served on a number of advisory boards including the Advisory Board of Kentucky Transportation Center at the University of Kentucky, Industry Advisory Board of Civil Engineering Department.

# Appendix B

# Spatial Scale of Road Effects
# on Ecological Conditions:
# Annotated Bibliography

Laurie Carr
TerraSystems Research

MAY 2003

## INTRODUCTION

The following is an annotated bibliography of road effects on ecological conditions, with a special emphasis on spatial scale. Only studies that directly measured the effect of roads on the surrounding environment were included. References were organized into two main categories, abiotic and biotic consequences. Within abiotic consequences, the effects of roads on hydrology, geomorphology, natural disturbances, and the effects of road chemicals on ecosystems are included. The biotic consequences category is further divided into three subcategories, genetic consequences, plant and wildlife population consequences and ecosystem consequences. Within each subcategory the effects of roads on structure, function and composition are included.

Every aspect of roads has some interaction with the surrounding environment, from road construction to maintenance. However, this list focuses on the effects generated from the presence and use of the road itself. With the exception of culverts, the impacts of road structures such as bridges or roadside lamps are not included. The reciprocal effects of the environment on roads are also not included.

For all ecological effects, references are organized by the scale of the study. Three scales were used: single segment, intermediate—political/ecological, and national. Single segment refers to studies that examine the impacts of a single road on an ecosystem. In this case, several roads may be involved in the study; however, the results do not address the cumulative effects of these roads. For example, the effect of road pollution on insect populations living adjacent to a road, or the barrier effect of a road on a species' movement would be classified as "single segment." Intermediate scale studies examine the *cumulative* effect of roads on a region. In other words, the combined effect of more than one road determines the results of the study. For example, the effect of several roads on one lake, range expansion by using roads as a dispersal corridor, or genetic isolation of a population surrounded by more than one road would be classified as intermediate. The boundary of a region is either politically (e.g., state of Florida, national park) or ecologically determined (e.g., a watershed, an animal's home range). National scale studies are very rare and they cover the effects of roads over an entire country.

The format for the bibliography is as follows:

## Sample Section

### Summary of Ecological Effects

A paragraph is written here explaining the ecological effects for the section. Ecological effects are italicized in paragraph.

Ecological effect:

**a. Single segment**
- Reference 1
- Reference 2
- Etc.

**b. Intermediate—-political/ecological**
- Reference 1
- Reference 2
- Etc.

**c. National**
- Reference 1

- Reference 2
- Etc.

## ABIOTIC CONSEQUENCES OF THE
## ECOLOGICAL EFFECT OF ROADS

### Hydrology and Geomorphology

### Summary of Ecological Effects

Roads have an impervious surface that collects and reroutes precipitation along its length or along roadside ditches. The temporary addition of waterflows from the road network affects flooding, groundwater supplies and channel morphology of *stream networks*. For logging roads specifically, precipitation and vehicular use results in *sediment production*. Sediment is carried into the watershed by wind or water and often contains chemicals. Roads that transect waterbodies cause *changes in waterflows* by restricting circulation. Roads can also become barriers to surface drainage when they are elevated compared to the landscape.

*Stream Networks*

### a. Single segment

- Almost all streams and intermittent channels crossed by a highway were channelized. Distance from 30 m to 400 m upslope and from 30 m to 500 m downslope of the road. Wetland drainage effects extended outward from the road for distances varying from 50 m to 500 m (Forman and Deblinger 2000).

### b. Intermediate—political/ecological

- When hard surface (road, buildings, parking areas) reaches 30-40% of the area about 30% of precipitation water becomes surface runoff, at 80-90% more than 55% of water becomes runoff. As a result, groundwater supplies may not be fully recharged, streams tend to degrade, and flooding often increases (Schueler 1995).
- Once hard-surface coverage exceeds 25%, streams in area tend to be degraded, as characterized by unstable channel morphology, polluted water, and highly altered or impoverished fish communities. In Seattle (USA) region, a road density of 5 km/km$^2$ (8 mi/mi$^2$) corre-

sponded to a hard-surface cover of about 20% in the watershed. Hard-surface model suggests stream networks are "impacted" at levels of hard surface as low as 10% coverage. (Center for Watershed Protection 1998).

- Results from a conceptual model of interactions between road networks and stream networks show that road networks appear to affect floods and debris flows, thus modifying disturbance patch dynamics in stream and riparian networks in mountain landscapes (Jones et al. 2000).

- Road effects on stream networks result in changes in watershed processes far removed from the site. Roads can increase drainage density by increasing waterflow from an impervious surface, and diverting subsurface water to surface and roadside ditches. This can result in floods, and alter aquatic and riparian ecological conditions, including fish populations in the lower parts of the stream systems (Eaglin and Hubert 1993).

- Stormwater flows along roads or ditches often effectively create new segments connected to the natural stream network. In a mountain forestry case approximately 57% of the road network functioned as an extension of the stream network, thereby increasing drainage density by 21% to 50% (Wemple, et al. 1996).

- Water from roadside ditches may be routed to streams, thus effectively increasing the density of stream channels in a watershed. A model predicted increases in the mean annual flood due to forest roads ranged from 2.2% to 9.5% (La March and Lettenmaier 2001).

**c.  National**
- No citations

*Sediment Production*

**a. Single segment**
- Inventories of almost 500 km of forest roads in several catchments indicate that untreated roads produced 1500 to 4700 $m^3$ of sediment per kilometer of road length (Madej 2001).

**b.  Intermediate—political/ecological**
- Lake Tahoe basin receives chemically laden sediment-bearing road runoff from a highly disturbed, road-laced watershed. Erosion along unpaved timber-harvesting roads caused by vehicular passage leads to sediment transported by wind or water (Zeigler et al. 2001).

• The greatest accumulation of fine sediments in streambeds was associated with logging road areas that exceeded 2.5% of the total basin area. Total road lengths of 2.5 km/km$^2$ of watershed basin produced sediment > 4 times natural rate (Cedarholm et al. 1981)

• A study of a forest road network in the western Cascade Range, Oregon found that debris slides from mobilized road fills were the dominant process of sediment production from roads. Overall, this study indicated that the nature of geomorphic processes influenced by roads is strongly conditioned by road location and construction practices, basin geology and storm characteristics (Wemple 2001).

### c. National
• No citations

*Changes in Waterflow*

### a. Single segment
• Restricted tidal flow in salt marsh of northern Massachusetts resulted in difference in salinity and hence vegetation of each side of culverts. Invasion of freshwater common reed (*Phragmites communis*) occurred in the side of low salinity (Massachusetts Executive Office of Environmental Affairs 1995).

• A causeway bisecting the Great Salt Lake in Utah altered circulation, resulting in changes in salt concentration between the two sides of the lake (Loving et al. 2000).

• A major effect of permanent roads in the artic is the blockage of surface drainage during spring snowmelt, thereby allowing water to accumulate beside a road in the relatively flat terrain. This can result in flooded areas or impoundments and sometimes road washouts (Walker et al. 1987).

### b. Intermediate—political/ecological
• Four causeways transect an estuary of Florida's Tampa Bay. This has altered circulation of the bay resulting in changes in contaminant and sediment transportation (Goodwin 1987).

### c. National
• No citations

# Chemical Characteristics

## Summary of Ecological Effects

Major sources of roadside pollution are vehicles, roads and bridges, and dry and wet (dust and rain) atmospheric deposition. Less frequent sources are accidental spills of oil, gasoline, and industrial chemicals. The majority of roadside chemicals come from vehicles (83%), 22% come from sanding and de-icing agents, 17% from roadbed and road surface wear, and 13% from herbicide and pesticide use. These figures do not include heavy metals and other chemicals that leach from bridges into streams and other water bodies (Federal Highway Administration 1996, Kobringer 1984).

Vehicular chemical pollutants include *mineral nutrients* (e.g. nitrogen and phosphorus), *heavy metals* (e.g. zinc and lead) and *organic* compounds (petroleum products). Of all the heavy metals, lead is the most studied ecologically. Lead was removed from gasoline in the mid-80's in North America, and lead levels in plants and animals are now relatively low (Forman, et al. 2003). The actual levels of heavy metals other than lead (Cadmium, copper, zinc, nickel, mercury and chromium) along roads, and their ecological effects remains poorly understood (Forman, et al. 2003). *De-icing salt* (sodium chloride) is applied on roads for snow and ice. Literature on the contamination of surface water and groundwater from road-salt is voluminous. Only a few examples are presented in this table.

*Mineral Nutrients*

### a. Single segment

• The largest source of phosphorus entering Lake Chocorua in New Hampshire was runoff from a multilane highway that passed near the eastern shoreline (Schloss 2002).

• Chemical nitrogen (from vehicle exhaust) enrichment of roadside soil favours few dominant flowering plants at the expense of more sensitive conifers, ferns, mosses, fungi, algae, and lichens in heathlands. This effect is higher along multilane highways compared to smaller roads (Angold 1997).

• Nitrogen (nitrogen oxides from traffic) caused increased growth in plant species in healthland communities in southern England.

Consequently species composition changed. Effects were measured up to 200 m (656 ft) from road (Angold 1992).

**b.  Intermediate—political/ecological**
- No citations

**c.  National**
- No citations

*Heavy Metals*

**a. Single segment**
- Increased levels of toxic heavy metals have been found up to 50 or 100 m (165 to 330 ft) of highways in the air, soil and plants (Ministry of Transport 1994b).
- Manganese concentrations in the soil along interstate highways in Utah are 100 times higher than historic levels. Roadside aquatic plants were sensitive bio-indicators of manganese contamination (Lytle et al. 1995).
- Some plants may have enhanced root growth as a result of soil contamination from roadside dust carrying trace metals (Wong et al. 1984).
- Combustion gases may cause reductions in species richness in arthropods, some groups flourished in the environment polluted by combustion gases. (Przybylski 1979).
- Overall lead levels were low in insects but high in earthworms, especially near highways (Marino et al. 1992).
- Early studies suggest that the abundance and diversity of invertebrates do not decline with increasing amount of metal pollution in roadside habitats (Muskett and Jones 1980).
- Lead contamination of insects near a highway in Kansas was more than 3 times level of those far from a road. Meadowlarks did not follow this trend reflecting little exposure to lead contaminated insects (Udevitz et al. 1980).
- Lead concentrations in little brown bats, short-tailed shrews, and meadow voles adjacent to a highway are high enough to cause mortality in domestic animals. (Clark 1979).
- Lead in roadside median strips of a highway was not considered a threat to adult ground-foraging songbirds (Grue et al. 1986).

**b.  Intermediate—political/ecological**
- No citations

**c.  National**
- No citations

*Organic*

**a.  Single segment**
- Green treefrog (*Hyla cinerea*) tadpole growth and metamorphosis was negatively associated with presence of petroleum contamination of freshwater, such as would occur from runoff (Mahaney 1994).

**b.  Intermediate—political/ecological**
- No citations

**c.  National**
- No citations

*De-icing Salt*

**a.  Single segment**
- Road-salt (CaCl) on an unpaved forest road inhibited crossing of the road by salamanders.  (Demaynadier and Hunter 1995).
- Survivorship of salamander species was lower in roadside pools that were heavily contaminated by de-icing salts.  (Turtle 2000).
- A highway that crosses the eastern portion of the Hubbard Brook Valley watershed has increased sodium and chloride concentrations in Mirror Lake (Likens et al. 1977).
- The gradual build-up of salt in soil and lowered moisture conditions can make it more difficult to maintain natural vegetation along roads (Thompson and Rutter 1986).
- Chloride concentrations in streams downstream from a salted highway were 31 times that upstream from the highway (Demers and Richard 1990.).

**b. Intermediate—political/ecological**
- De-icing salt used in the Rochester, New York, area dissolved and entered Lake Ontario via storm water drainage. This caused increased salinity of Irondequoit Bay to the extent that it prevented vertical mixing of the bay water in spring (Bubeck et al. 1971).
- Road salting caused changes in salt concentration of a small meromictic lake. This weakened the lake's meromictic (chemical-dependent) stability and had implications for primary productivity of the lake (Kjensmo 1997).

**c. National**
- No citations

**Natural Disturbance**

**Summary of Ecological Effects**

Logging and mountain roads are susceptible for creating landslides due to unstable soil, steep slopes and high road densities (Havlick 2002). Roads can increase water discharge rates in a watershed increasing the potential for landslides.

*Landslides*

**a. Single segment**
- No citations

**b. Intermediate—political/ecological**
- Eighty-eight percent of landslides in Boise and Clearwater national forests in Idaho were road related. Most landslides on Idaho's South Fork of the Salmon River were also road related (Megahan 1980).
- After the 1964 flood in the Pacific Northwest, landslide frequency due to forest roads was up to 30 times the rates in unmanaged forested areas (Swanson and Dryness 1975).

**c. National**
- No citations

# GENETIC CONSEQUENCES OF THE
# ECOLOGICAL EFFECT OF ROADS

## Structure

### Summary of Ecological Effects

Roads can act as a *barrier to movement* through road-kill and be-havioralioural road avoidance. As barriers, roads subdivide continuous populations and reduce gene flow between sub-populations. The result is isolated and smaller populations. Genetic consequences for larger species with smaller population sizes or for endangered species with low dispersal abilities are thought to be greater (Gerlach and Musolf 2000) but have not been investigated.

*Barrier to Movement*

### a.  Single segment
  • A recent highway, at least 25 yrs old, was shown to have an effect on the genetic sub-structuring of bank vole (*Clethrionomys glareolus*) populations due to reduced gene flow. Small population size, a country road, and a railway did not appear to affect genetic structure (Gerlach and Musolf 2000).

### b.  Intermediate—-political/ecological
  • Common frog (*Rana temporaria*) population surrounded by roads, a highway and a railway had reduced average amount of heterozygosity (genetic variation) and genetic polymorphism (diversity of forms) (Reh and Seitz 1990).

### c.   National
  • No citations

## Function

### Summary of Ecological Effects

*Isolated populations* have a lower chance of survival without the demographic and genetic input of immigrants, and of recolonization after extinction (Lande 1988). Low gene flow in isolated populations has the negative result of increased inbreeding (weak offspring) and decreased fecundity therefore increasing the probability of extinction (Hartl et al. 1992 as cited in Gerlach and Musolf 2000). Little is known on the long-term ecological effects of roads on mammalian speciation through isolation (Baker 1998).

*Isolated Populations*

**a.** **Single segment**
- No citations

**b.** **Intermediate—-political/ecological**
- No citations

**c.** **National**
- No citations

## Composition

### Summary of Ecological Effects

Roads have a *filtering effect*, in that different species or segments of a population may interact distinctively in relation to roads. This can result in natural selection of specific genes.

*Filtering Effect*

### a. Single segment
- Selection for early flowering and salt tolerance has developed in populations of *Anthoxanthum odoratum* L., growing along a roadside and in adjacent pastures, in less than 40 years (Kiang 1982).
- Genetic immunity to car pollution has selected for micropopulations of *Tenebrionidae* that have lived along roadsides for many generations. Beetle larvae that did not posses the immunity perished at instar II due to pollution exposure (Minoranskii and Kuzina 1984).

**b.   Intermediate—-political/ecological**
*   No citations

**c.   National**
*   No citations

## CONSEQUENCES OF THE ECOLOGICAL EFFECT OF ROADS ON SPECIES AND POPULATIONS

### Structure

### Summary of Ecological Effects

Wildlife-road interactions (avoidance, mortality, use of roadside habitats) may be regulated by the age or sex of the individual. Differences in mortality and use of the landscape can result in shifts in *wildlife population structure*. No studies were found on plant population structure as influenced by roads.

*Wildlife Population Structure*

### a.  Single segment

*   Roadside territories are sinks for Florida scrub jays (*Aphelocoma coeulescens*) populations. Vehicular mortality along a two-lane highway was significantly higher for new immigrants, the rate dropped after 3 years of living in a road territory. Fledglings also experienced significantly higher mortality compared to adults (Mumme et al. 2000).
*   Road mortality is indicated in a shift in age structure toward younger age classes for painted turtles (*Chrysemys picta*) populations. A higher percentage of adult turtles were found farther from a segment of Route 93 adjacent to the Ninepipe National Wildlife Refuge in Mission Valley, Montana (Fowle 1990).
*   The four-lane, divided Trans-Canada Highway segment within Banff National Park in Alberta, served as a complete barrier against the movement of adult female bears (*Ursus arctos*) and as a partial filter-barrier for adult males (Gibeau 2000).
*   Radio-tracking showed that female stoats inhabiting beech forests avoided the road in study area but males preferred it (Murphy and Dowding 1994).

**b.** **Intermediate—-political/ecological**
- No citations

**c.** **National**
- No citations

## Function

### Summary of Ecological Effects

Roads and roadside vegetation increase the diversity of habitats in the landscape by providing *additional habitat* sources. Some species are able to take advantage of the foraging and habitat opportunities presented by roads. In some managed landscapes, roadsides have been left as vestigial strips of original habitat. These vegetated strips, or road reserves, become refuges for animal species that depend on natural communities that have been reduced in the landscape. The benefits provided by roads must however, be weighed against the increased probability of those species incurring vehicular collisions or hunting. On the other hand, roadsides and adjacent land have a *reduced habitat quality* for many species due to disturbance from traffic presence and noise.

Roads and roadsides can contribute to the spread of wildlife and plants by providing suitable habitat and a highly connected *dispersal corridor*. While this connectivity has the potential to be a dispersal corridor for wildlife, there is little existing evidence for that function at the population level (Forman et al. 2003).

Roads can also present as a complete or partial *movement barrier* to wildlife, through behavioral avoidance or a high mortality rate. Species that exhibit a behavioral avoidance of roads are less likely to incur mortality. However they are vulnerable to a loss of landscape connectivity especially if they have multiple resource needs or large territories. Road avoidance also results in the alteration of habitat use such that the *distribution* of some species conforms to the presence of roads.

### *Additional Habitat*

#### a. Single segment
- An extensive study along interstate highways in the U.S. showed that grassland and several generalist small mammal species were

more abundant on verge habitat than on adjacent habitat (Adams and Geis 1983).

- In median strips, small mammal density tended to be highest where unmowed grassy roadsides bordered wooded strips (Adams 1984).

- At six interchanges along Highway 417 in Ottawa, Canada, the density of woodchucks (*Marmota monax*) per hectare exceeded any density previously reported for this species in any habitat (Woodard 1990).

- Road verges with hedges and tall grass provide habitat for vole and mice species (Bellamy et al. 2000).

- The abundance of the generalist *Melomys cervinipes* increased near a rainforest road, while *Rattus* sp. decreased. *Rattus* sp. prefer undisturbed habitat (Goosem 2000).

- White-tailed deer (*Odocoileus virginianus*) graze in road rights-of-way in forested habitat (Carbaugh et al. 1975).

- Raptors have higher use of roadsides compared to adjacent habitat because of the greater availability of perch sites and wide roadsides (Meunier et al. 2000).

- Roadside ditches act as additional vernal pool habitats for aquatic invertebrates in the Central Valley of California (Koford 1993).

- Road reserves provide habitat for uncommon native species when juxtaposed in disturbed landscapes. Roadsides of one multilane highway in the Netherlands were composed of 20% of uncommon species. In this case, roadsides represented more suitable habitat then the surrounding landscape (Ministry of Transport 2000, Sykora et al. 1993).

- Numbers of nests and grassland passerine species increased with roadside width along rural interstate (four-lane) and secondary rights-of-way in central Illinois. Expansion of row-crop farming means roadsides critical for sustaining birds that nest in edges and ecotones (Warner 1992).

- Species richness and abundance of habitat-sensitive butterflies increased on roadsides restored to native prairie vegetation compared to grassy and weedy roadsides (Ries et al. 2001).

## b. Intermediate—-political/ecological

- Raven (*Corvus corax*) numbers showed a positive relationship with increasing number of linear rights-of-way which ran in parallel. Possible reasons are increase in edge habitat and carrion along road (Knight et al. 1995).

**c. National**
- A survey of 212, 220 ha of roadside habitat in the U.K. found that road verges provided habitat for a large portion of the native plants, mammals, reptile and invertebrates. A number of nationally rare plant species also depended on roadside habitats for survival. Verges were mostly grassland (Way 1977).

*Reduced Habitat Quality*

**a. Single segment**
- A series of seminal papers that showed that of 43 species of woodland breeding birds, 26 species (60%) showed reduced densities near highways. Traffic noise explained the most variation in bird density in relation to roads in a regression model. (Reijnen and Foppen 1994). This effect also occurred for grassland birds (Reijnen et al. 1996), and is more important in years with a low overall population size (Reijnen and Foppen 1995).
- High traffic volume had significant effects on presence and breeding of grassland birds in roadsides in an outer suburban landscape of Massachusetts. Lower traffic volumes had no effect (Forman et al. 2002).
- Dutch studies of four bird species in open grasslands found the traffic disturbance effect on population density extended a further distance next to a busy highway compared to a rural road (van der Zande et al. 1980).
- Dispersal of breeding male willow warblers (*Phylloscopus trochilus*) was actively directed away from road, constituting an "escape" from low quality roadside habitat (Foppen and Reijnen 1994).
- Population density of the horned lark (*Eremophila alpestris*) increased with distance from roads in an agricultural landscape (Clark and Karr 1979).
- Flocks of pink-footed geese (*Anser brachyrhynchus*) and graylag geese (*A. Anser*) were not found within 100 m (328 ft) of the nearest road. Geese also never visited fields with centers closer than 100 m from roads (Keller 1991).
- Ovenbird (*Seiurus aurocapillus*) territory size decreased with distant from unpaved roads in a forested region in Vermont. Habitat quality for ovenbirds may be lower within 150 m (492 ft) of roads, thus requiring a larger area for foraging, and possibly reducing reproductive success (Ortega and Capen 1999).

- In northern Maine total population of breeding birds near a highway was not significantly different than at a greater distance. However, the total population was less than half as equal area of forest habitat and population densities were 79% of those in forest habitat (Ferris 1979).
- Proximity of Florida scrub jay territories to paved road with approximately 500-750 vehicles per day had no effect on nesting success (Beatley 2000).

**b. Intermediate—-political/ecological**
- Analysis of the total effect of The Netherlands's most dense network of main roads on 'meadow birds' shows a possible population decrease of 16%. Population decrease was attributed to reduced habitat quality and traffic noise in previous studies (Reijnen et al. 1997).

**c. National**
- No citations

*Dispersal Corridor*

**a. Single segment**
- Roadside vole species (*Microtus*) that could not move through villages and towns was able to expand its range 90 km (56 mi) along an Illinois multilane highway with continuous vegetation (Getz et al. 1978).
- Large mammals such as wolves and lynx have been noted following roads and trails that have little human travel (Thurber et al. 1994).
- Arthropods (carabid beetles and lycosid spiders) exhibited longitudinal movements along a roadside and reduced crossing rates for paved and gravel roads (Mader 1990, Vermeulen 1994).

**b. Intermediate—-political/ecological**
- Pocket gophers (*Thomomys bottae*) were able to extend their range along roadside verges that provided suitable habitat (Huey 1941).
- Grassland plants use road and railway corridors for dispersal (Tikka et al. 2001).
- Ravens feeding on road-killed animals along roadsides move through landscape (Knight and Kawashima 1993).
- Cane toads (*Bufo marinus*), an introduced species in Australia, use roads as dispersal corridors, especially in dense vegetation. This be-

havior allowed the toads to extend their range into otherwise inaccessible areas (Seabrook and Dettmann 1996).

• In North Dakota, the harvester ant (*Pogonomyrex occidentalis*) has expanded its range several kilometers in 12 years, by moving along roadsides (Demers 1993).

• Study in Kakadu National Park, Australia suggests that tourist vehicles were partly responsible for the spread of weed species in the park (Lonsdale and Lane 1994).

• Roadside ditches are a method of dispersal for water-resistant seeds. An escaped ornamental, purple loosestrife (*Lythrum salicaria*) has invaded roadsides in New York State and formed conspicuous monocultures along ditches and culverts (Wilcox 1989).

**c. National**
• No citations

*Movement Barrier*

**a. Single segment**
• Small mammal crossing seems to be highly dependent of road width. Small mammals were found to be reluctant to cross roads greater than 20 m (66 ft) (Oxley et al. 1974, Richardson et al. 1997). This also holds true for *Rattus* sp. on rainforest roads, but only during breeding season (Goosem 2001). Road crossings by small ground mammals in Australia were inversely related to road width (Barnett et al. 1978). In Kansas, small road clearances less than 3 m (10 ft) have been shown to effect small mammals such as voles and rats (Swihart and Slade 1984).

• Restricted movements of small mammals across roads are probably due to behavior rather than an inability to cross the road. A forest road did not affect movement of yellow-necked mice (*Apodemus flavicollis*), but restricted that of bank voles (*Clethrionomys glareolus*) (Bakowski and Kozakiewicz 1988). An experimental study showed rodents in the Mojave Desert hesitate to cross roads even though they may travel long distances. (Garland and Bradley 1984) Rodents would cross highways in southwestern Texas when the habitats on opposing sides were similar (Kozel and Fleharty 1979).

• In Ontario, movements of mice across a lightly traveled road of 6-15 m (20-49 ft) wide had a probability of less than 10% compared to movements in the adjacent habitats (Merriam et al. 1989).

• Mobility diagrams show significant isolation effects of roads on populations of forest-dwelling mice (*Apodemus flavicollis*) and carabid beetles, neither of which crossed highways or forest roads (Mader 1984).

• While a road inhibited movements of small rainforest mammals, genetic isolation of populations separated by a road less than 12 m wide is slight (Burnett 1992).

• Mixed-species flocks use the vegetation along the edges of a narrow, rarely used road through otherwise undisturbed Amazonian forest road, but because they are unwilling to cross the open area, the road becomes a flock territory boundary (Develey and Stouffer 2001).

• Road crossings may act as barriers to fish migration. The movement of fish through crossings was inversely proportional to the water velocity (Warren and Pardew 1998).

• Steep slopes and low roughness coefficients of culverts frequently cause high water velocities that have prevented passage of Arctic grayling, long-nose suckers, northern pike, salmon and trout (Derksen 1980, Kay and Lewis 1970, as cited in Belford and Gould 1989) (Belford and Gould 1989).

• Salamander abundance was 2.3 times higher at forest interior vs. roadside sites. A large road was a partial barrier depending on type of movement (natal dispersal, migratory, home-range) (Demaynadier and Hunter 2000).

• Road-forest edges significantly hindered amphibian movements in a forest tract in southern Connecticut (Gibbs 1998).

• Arthropods (carabid beetles and lycosid spiders) exhibited longitudinal movements along the roadside and reduced crossing rates for paved and gravel roads (Mader et al. 1990).

• Bumblebees (*Cephalanthus occidentalis* L.) showed high site fidelity and only rarely crossed roads or railroads (Bhattacharya et al. 2003).

• Mark-recapture study showed almost no land snails (*Arianta arbustorum*) crossed road in central Sweden. Results suggest that snail populations separated by paved roads with high traffic densities may be isolated from each other. (Baur and Baur 1990).

## b. Intermediate—-political/ecological

• Caribou (*Rangifer tarandus*) density is inversely related to road density. At 0.3 km/km$^2$ (0.5 mi/mi$^2$), the population declined by 63%. At 0.9 km/km$^2$ (1.4 mi/mi$^2$), it was 86% lower. Areas with over

0.9 km/km$^2$ of road length have no cow-calf pairs, and maternal females show a displacement of 4 km (2.5 mi) away from roads (Nellemann and Cameron 1996). The greatest avoidance zone or effect-distance reported is 5 km (3 mi) for caribou (Nellemann and Cameron 1998).

- An increase in the number of roads and railways and traffic volume has subdivided the original caribou population of Scandinavia into 26 separate herds. Areas with 5 km (3 mi) of roads and power lines are essentially avoided. Due to continued development the survival of these populations is in doubt (Nellemanna et al. 2001).
- Cumulative effect of parallel linear developments may affect movement of caribou. They were found to cross a road and a control area with similar frequency, but when a pipeline ran parallel to a road, crossing frequencies dropped significantly (Curatolo and Murphy 1986).

**c.  National**
- No citations

*Distribution*

**a.  Single segment**
- Deer and elk avoid roads in winter ranges in Colorado, particularly areas within 200 m of a road (Rost and Bailey 1979).
- Grizzly and black bears (*U. americanus*) used habitat < 1 km from a road or trail less than expected in the Cabinet Mountains, Montana (Kasworm and Manley 1990).

**b.  Intermediate—-political/ecological**
- Data indicates an increased survival rate, a reduction in movements, and home range size of Roosevelt elk (*C. elaphus roosevelti*) during a period of limited vehicular access (Cole et al. 1997).
- As road density increases from 2 to 3 mi/mi$^2$ (1.2-1.9 km/km$^2$) to 5 to 6 mi/mi$^2$ (3.1-3.7 km/km$^2$), elk use of habitat declines by approximately 50% (Lyon 1983.).
- Black bears shift the locations of their home ranges when faced with increases in road densities in areas of previously lower road densities. (Brody and Pelton 1989).
- Grizzly bears in the Swan Mountains of Montana used habitats with lower total road density and avoided buffer areas surrounding roads having >10 vehicles per day (Mace et al. 1996).

- Road density explains the juxtaposition of home ranges and distribution of bobcats (*Felis rufus*) in northern Wisconsin (Lovallo and Anderson 1996).

### c. National
- No citations

## Composition

### Summary of Ecological Effects

Plant *species richness* in roadsides depends on roadside management (mowing, planting, amount of disturbance and pollution). Plant communities change over time depending on the successional pattern altered by roadside management. Wildlife behavior, road mortality and reaction to pollutants determine wildlife species richness. *Road mortality* has a dual impact on regional populations through a direct loss of individuals and an indirect decrease in landscape connectivity. Experimentation to separate these two effects has not been conducted and emphasis of road mortality remains on the loss of individuals. Few studies compare the amount of road mortality to other sources of mortality or population size. This results in little being known on the population effects of road mortality.

There is widespread concern over the invasion of *non-native plants in roadsides and adjacent landscapes*. Roadsides can be ideal habitats for non-native or exotic species. Increased light, soil disturbance, variable soil moisture conditions and dispersal vectors (vehicles, connected habitat) have resulted in an abundance of non-native species in our roadsides. Non-natives are introduced into roadsides via intentional plantings to reduce erosion, from travelers (human and animal) and escaped ornamentals.

*Species Richness*

### a. Single segment
- Wider, more open unpaved roads tended to produce steeper declines in roadside abundance and richness of macroinvertebrate soil fauna, and leaf-litter depth. Effects persisted up to 100 m into the Cherokee National Forest, USA (Haskell 2000).

• Mowing twice a year, early and late in the growing season, created highest plant diversity, in wide roadsides beside multilane highways. Mowing once a year either early or late in the season or mowing every other year early in the season resulted in fewer species. The lowest diversity was found with no mowing (Ministry of Transport 1994a).

• Based on sampling 50 km (31 miles) of roadsides in South Island, New Zealand, total number of non-natives was not significantly different from the road shoulder, to the outer roadside. However, native species diversity increased over a gradient from inner to outer roadside (Ullmann et al. 1998).

### b. Intermediate—-political/ecological

• Significant effects of road density within 2 km (1 mi) of wetlands have been noted for wetland species richness of reptiles, amphibians, birds and plants (Findlay and Houlahan 1997). Although a subsequent study found herptile, vascular plants, and bird species richness in wetlands was more accurately described by road density 30 to 40 years ago (Findlay and Bourdages 2000).

• Only areas where road density is less than 0.72 km/km$^2$ (1.16 mi/mi$^2$) seem to support vibrant populations of wolves (*Canis lupus*) in Minnesota (Fuller 1989, Mech et al. 1988), Wisconsin (Mladenoff et al. 1999, Theil 1985), the western part of the Great Lakes region of the USA (Mladenoff et al. 1995), and Ontario (Canada) (Jensen et al. 1986). An exception to the trend is an established wolf population in a fragmented area of Minnesota with a road density of 1.42 km/km$^2$ (2.29 mi/mi$^2$) (Merrill 2000).

• Paved and unpaved roads and power line rights-of way were examined for their effects on the relative abundance and community composition of forest-nesting birds in southern New Jersey. At a landscape scale, small reductions in forest area because of corridors may be cumulatively significant for forest-interior birds (Rich et al. 1994).

### c. National

• No citations

*Road Mortality*

### a. Single segment

• Results suggest roads and traffic are likely to reduce hedgehog density by about 30%, which may affect the survival probability of local populations. Hedgehog population density was estimated at 15 areas near a road (Huijser and Bergers 2000).

• Road mortality resulting from increase in traffic speed along a road was suspected in the local extinction of eastern quolls and 50% reduction of Tasmanian devils (Jones 2000).

• Two taxa of special conservation interest, the Mexican rosy boa (*Lichanura trivirgata*) and the Organ Pipe shovelnosed snake (*Chionactis palarostris*), appear to be relatively strongly impacted by highway mortality. Estimate close to 4000 snakes killed on a stretch of road (Rosen and Lowe 1994).

• A study of painted turtles at varying distance from roads found both lower density and higher mortality near roads. A total of 205 painted turtles (*Chrysemys picta*) were killed along a 7.2 km (4.5 mi) road segment that bisects a series of prairie pothole (Fowle 1990).

• Road mortality had a significant effect on local densities of amphibians along a two-lane road in Ontario, Canada. The density of frogs and toads in the roadside and adjacent habitat decreased with more traffic and higher road mortality rates (Fahrig et al. 1995).

• Results estimate that 10% of the adult population of *Pelobates fuscus* and brown frogs (*Rana temporaria* and *R. arvalis*) are killed annually by traffic on a two-lane road in Northern Denmark. Population sizes were estimated for all ponds within 250 m of the road (Hels and Buchwald 2001).

• Vehicular mortality was insufficient to affect roadside butterfly populations along 12 main roads in the U.K. (Munguira and Thomas 1992).

• Traffic volume was related to badger (*Meles meles*) deaths in southwest England. Above a certain level, all high traffic volume roads had 6 times greater mortality then low volume roads. Results suggest high volume roads may discourage badgers crossing roads (Clarke et al. 1998).

## b. Intermediate—-political/ecological

• The number of roads and road density was the strongest determinant in the decline of badgers in the Netherlands (*Meles meles*), likely due to traffic mortality (Van Der Zee et al. 1992).

• Vehicular collisions from the roads in a national park in southwestern Spain are one of the largest sources of mortality for Iberian lynx (*Lynx pardalis*) (Ferreras et al. 1992).

• Vehicle collisions along the highways and gravel roads in the Kenai National Wildlife Refuge in Alaska are reported as the largest source of mortality for radio-collared female moose (*Alces alces*) (Bangs et al. 1989).

• Road density (corrected for traffic volume) within 0.75 km (0.5 mi) from 109 Dutch moorland pools significantly explained the probability as to whether or not the moor frog were present in a pool (Vos and Chardon 1998).

• Total traffic volume (dependent on road density) had a significant negative effect on the population density of leopard frogs (*Rana pipiens*) within a radius of 1.5 km of ponds. Traffic volume had no effect on the less mobile species green frogs (*Rana clamitans*) (Carr and Fahrig 2001).

• Results of a spatially explicit individual-based model reveal a much stronger impact of road mortality than the barrier effect on wildlife populations. The influence of traffic mortality is always much more significant when the proportions of individuals avoiding the road and those that are killed on the road (in relation to the number of individuals encountering roads) in the two situations compared are the same. Results were examined for several road network configurations (Jaeger and Fahrig 2001).

**c.   National**
   • No citations

*Non-native Plants in Roadsides and Adjacent Landscapes*

**a.  Single segment**
   • Nearly every sampling unit along abandoned (low disturbance and light levels), low- and high-use (highest disturbance and light levels) roads in the H.J. Andrews Experimental forest contained at least one exotic plant species. Exotic species were significantly more frequent under high or medium light than under low light conditions. Roads and streams apparently serve as, corridors for dispersal, suitable habitat, and propagule reservoirs (Parendes and Jones 2000).

   • Alien plant species from an old road have spread into disturbed sites in sclerophyll forest communities, Australia. Frequency of plants declined with increasing distance from road and was correlated with the reduction in diffuse light (Amor and Stevens 1976).

   • Introduction of limestone rock material for roadbeds changed the soil attributes of the roadside in a sandy Florida scrub area, which supported non-indigenous species (Greenberg et al. 1997).

**b. Intermediate—-political/ecological**

- Transects through 13 of the lower Florida Keys found that areas closest to paved roads, with the largest amount of development within a 150 m radii, had the highest probability of presence of predatory, non-native, red imported fire ants (*Solenopsis invicta*) (Forysa et al. 2002).

- Non-native invasive multiflora rose (*Rosa multiflora*) was planted along highways in Virginia and elsewhere and quickly spread into natural habitats to become a widespread infestation (Stiles 1980).

**c.    National**

- No citations

# CONSEQUENCES OF THE ECOLOGICAL EFFECT OF ROADS ON ECOSYSTEMS

## Structure

### Summary of Ecological Effects

Although roads are relatively narrow disturbances, they have create a disproportionate amount landscape *fragmentation*.  Roads create an abundance of edge habitat, reducing habitat for organisms adapted to interior conditions.  Roads also fragment populations when they become a barrier to wildlife movement.

*Fragmentation*

**a.    Single segment**

- No citations

**b. Intermediate—-political/ecological**

- Fragmentation was quantified due to roads in 30 areas of a 213-ha section of Medicine Bow-Route National Forest in southeastern Wyoming.  Average road density was 2.52 km/km$^2$.  Roads created 1.54-1.98 times the edge habitat created by clearcuts and the total landscape area affected was 2.5-3.5 times the actual area occupied by roads (Reed et al. 1996).

• When stand boundaries were delineated by roads, there was a large increase in small stands with simple shapes, concurrent with a decline in the number of stands > 100 ha. Early-seral forest were in greater proportion adjacent to roads, suggesting effect of roads on landscape structure was somewhat localized (Miller et al. 1996).

• For eight land cover types in the Northern Great Lakes Region, USA, 5% to 60% of a land cover type was affected by roads, depending on the assumed depth-of-edge influence (DEI). Roads increased number of patches and patch density, and decreased mean patch size and largest patch index (Saunders et al. 2002)

**c.  National**
   • No citations

**Function**

**Summary of Ecological Effects**

*Pollutants* from roads and vehicles can alter ecological processes by weakening defense mechanisms in plants, increasing available nutrients and bio-accumulating in the food chain.

*Pollutants*

**a.  Single segment**
   • High nitrogen content of roadside vegetation along a busy motorway in the United Kingdom was believed responsible for the rate of increase and the outbreaks in insect populations. Grasshoper density (*Chorthippus brunneus*) was higher on motorway sites compared to sites away from the motorway (Port and Thompson 1980).

   • De-icing salts may cause physiological stress in roadside trees in Switzerland, making them more susceptible to aphid infestations (Braun and Fluckiger 1984).

   • Salt in roadside soils may increase available nitrogen and some minerals. This is a possible cause for *Lolium perenne* growing more vigorously in soil from near roads (Spencer and Port 1988).

- Some invertebrate groups appeared to increase with proximity to certain major roads sampled. Higher levels of lead found in small mammals at the higher trophic level (Goldsmith and Scanlon 1977).
- Lead levels, and not zinc, were higher in invertebrates in close proximity to a road. However, the levels of heavy metals are too low to be toxic in their animal predators (Wade et al. 1980).

**b.  Intermediate—-political/ecological**
- No citations

**c.  National**
- No citations

## Composition

### Summary of Ecological Effects

*Environmental characteristics* such as soil moisture, air temperature, soil compaction and composition, and light differ markedly on roads compared to adjacent surfaces. Dust from road surfaces that is carried onto adjacent surfaces extends the environmental effect of roads. These changes are translated into changes of species composition, and in tundra landscapes, pre-mature thawing of permafrost.

*Environmental Characteristics*

**a. Single segment**
- Vehicle tracks in northern Alaska have higher temperatures, deeper thaw of permafrost, and higher concentrations of soil phosphate compared to undisturbed tundra. Tracks also have fewer species of plants, more evergreen shrubs, and greater dominance by grass-like species (Chapin and Shaver 1981).
- Heavy dusting near roads altered depth of thaw layer in ground, compared with areas far from roads. Total plant biomass was lower, and plant communities were species-impoverished, possibly due to increased soil acidity caused by road dust (Auerbach et al. 1997).
- Dust from gravel roads reflects incoming solar heat, which in turn is carried by wind into adjoining natural ecosystems, thus accelerat-

ing thermokarst (melting of the permafrost). In some situations, spring snowmelt can occur as much as two weeks earlier, and extend up to 100 m (328 ft) outward, near heavily traveled gravel roads (Walker et al. 1987).

**b.    Intermediate—-political/ecological**
*   No citations

**c.    National**
*   No citations

## BIBLIOGRAPHY

Adams, L. W., and A. D. Geis. 1983. Effects of Roads on Small Mammals. *Journal of Applied Ecology* 20:403-15.

Adams, L. W. 1984. Small Mammal Use of an Interstate Highway Median Strip. *Journal of Applied Ecology* 21:175-78.

Amor, R. L., and P. L. Stevens. 1976. Spread of Weeds from a Roadside into Sclerophyll Forests at Dartmouth, Australia. *Weed Research* 16:111-18.

Angold, P. G. 1992. *The Role of Buffer Zones in the Conservation of Semi-Natural Habitats.* Ph.D. University of Southhampton.

Angold, P. G. 1997. The Impact of a Road Upon Adjacent Heathland Vegetation —Effects on Plant Species Composition. *Journal of Applied Ecology* 34:409-17.

Auerbach, N., M. D. Walker, and D. A. Walker. 1997. Effects of Roadside Disturbance on Substrate and Vegetation Properties in Arctic Tundra. *Ecological Applications* 7:218-35.

Baker, R. H. 1998. Are Man-Made Barriers Influencing Mammalian Speciation? *Journal of Mammalogy* 79:730-71.

Bakowski, C., and M. Kozakiewicz. 1988. The Effect of Forest Road on Bank Vole and Yellow-Necked Mouse Populations. *Acta Theriologica* 33:345-53.

Bangs, E. E., T. N. Bailey, and M. R. Portner. 1989. Survival Rates of Adult Female Moose on the Kenai Peninsula, Alaska. *Journal of Wildlife Management* 53:557-63.

Barnett, J. L., R. A. How, and W. F. Humphreys. 1978. The Use of Habitat Components by Small Mammals in Eastern Australia. *Australian Journal of Ecology* 3:277-85.

Baur, A., and B. Baur. 1990. Are Roads Barriers to Dispersal in the Land Snail *Arianta Arbustorum*? *Canadian Journal of Zoology* 68:613-17.

Beatley, T. 2000. *Green Urbanism: Learning from European Cities.* Island Press, Washington, D.C.

Belford, D. A., and W. R. Gould. 1989. An Evaluation of Trout Passage through Six Highway Culverts in Montana. *North American Journal of Fisheries Management* 9:437-45.

Bellamy, P. E., R. F. Shore, D. Ardeshir, J. R. Treweek, and T. H. Sparks. 2000. Road Verges as Habitat for Small Mammals in Britain. *Mammal Review* 30:131-39.

Bennet. 1991. Roads, Roadsides and Wildlife Conservation: A Review. Pages 99-118 *in* Saunders and Hobbs, editors. *Nature Conservation 2: The Role of Corridors.* Surrey Beatty and Sons.

Bhattacharya, M., R. B. Primack, and J. Gerwein. 2003. Are Roads and Railroads Barriers to Bumblebee Movement in a Temperate Suburban Conservation Area? *Biological Conservation* 109:37-45.

Braun, S., and W. Fluckiger. 1984. Increased Population of the Aphis *Aphis Pomi* at a Motorway. Part 2—the Effect of Drought and Deicing Salt. *Environmental Pollution (Series A)* 36:261-70.

Brody, A., and M. Pelton. 1989. Effects of Roads on Black Bear Movements in Western North Carolina. *Wildlife Society Bulletin* 17:5-10.

Bubeck, R. C., W. H. Diment, B. L. Deck, A. L. Baldwin, and S. D. Lipton. 1971. Runoff of Deicing Salt: Effect on Irondequoit Bay, Rochester, New York. *Science* 172:1128-32.

Burnett, S. E. 1992. Effects of a Rainforest Road on Movements of Small Mammals: Mechanisms and Implications. *Wildlife Research* 19:94-104.

Carbaugh, B., J. P. Vaughan, E. D. Bellis, and H. B. Graves. 1975. Distribution and Activity of White-Tailed Deer Along an Interstate Highway. *Journal of Wildlife Management* 39:570-81.

Carr, L. W., and L. Fahrig. 2001. Effect of Road Traffic on Two Amphibian Species of Differing Vagility. *Conservation Biology* 15:1071-78.

Cedarholm, C. J., L. M. Reid, and E. O. Salo. 1981. Cumulative Effects of Logging Road Sediment on Salmonid Populations in the Clearwater River, Jefferson County, Washington. Pages 38-74 *in Washington Water Research Council. 1981. Proceedings from the Conference on Salmon-Spawning Gravel: A Renewable Resource in the Pacific Northwest.* Washington State University, Washington Water Research Centre, Report 39, Pullman, Washington.

Center for Watershed Protection. 1998. *Rapid Watershed Planning Handbook: A Comprehensive Guide for Managing Urbanizing Watersheds.* Center for Watershed Protection, Elliot City, Md.

Chapin, F. S., and G. R. Shaver. 1981. Changes in Soil Properties and Vegetation Following Disturbance of Alaskan Arctic Tundra. *Journal of Applied Ecology* 18:605-17.

Clark, D. R. 1979. Lead Concentrations: Bats vs. Terrestrial Small Mammals Collected near a Major Highway. *Environmental Science and Technology* 13:338-40.

Clark, W. D., and J. R. Karr. 1979. Effects of Highways on Red-Winged Black-bird and Horned Lark Populations. *Wilson Bulletin* 91:143-45.

Clarke, G. P., P. C. L. White, and S. Harris. 1998. Effects of Roads on Badger *Meles Meles* Populations in South-West England. *Biological Conservation* 86:117-24.

Cole, E. K., M. D. Pope, and R. G. Anthony. 1997. Effects of Road Manage-ment on Movement and Survival of Roosevelt Elk. *Journal of Wildlife Managemen.* 61:1115-26.

Curatolo, J. A., and S. M. Murphy. 1986. The Effects of Pipelines, Roads and Traffic on the Movements of Caribou, *Rangifer Tarandus. Canadian Field Naturalist* 100:218-24.

Demaynadier, P. G., and M. L. Hunter. 1995. The Relationship between Forest Management and Amphibian Ecology: A Review of the North American Literature. *Environmental Review* 3:230-61.

Demaynadier, P. G., and M. L. Hunter. 2000. Road Effects on Amphibian Movements in a Forested Landscape. *Natural Areas Journal* 20:56-65.

Demers, C. L., and W. S. Richard. 1990. Effects of Road Deicing Salt on Chlo-ride Levels in Four Adirondack Streams. *Water, Air, Soil Pollution* 49:369-73.

Demers, M. N. 1993. Roadside Ditches as Corridors for Range Expansion of the Western Harvester Ant (*Pogonomyrmex Occidentalis* Cresson). *Land-scape Ecology* 8:93-102.

Develey, P. F., and P. C. Stouffer. 2001. Effects of Roads on Movements by Understory Birds in Mixed-Species Flocks in Central Amazonian Brazil. *Conservation Biology* 15:1416

Eaglin, G. S., and W. A. Hubert. 1993. Effects of Logging and Roads on Sub-strate and Trout in Streams of the Medicine Bow National Forest, Wyo-ming. *North American Journal of Fisheries Management* 13:844-46.

Fahrig, L., J. H. Pedlar, S. E. Pope, P. D. Taylor, and J. F. Wegner. 1995. Effect of Road Traffic on Amphibian Density. *Biological Conservation* 74: 177-82.

Federal Highway Administration. 1996. *Evaluation and Management of High-way Runoff Water Quality.* FHWA-PD-96-032, U.S. Department of trans-portation, Washington, D.C.

Ferreras, P., J. J. Aldama, J. F. Beltran, and M. Delibes. 1992. Rates and Causes of Mortality in a Fragmented Population of Iberian Lynx—*Felis Pardina* Temminck, 1824. *Biological Conservation* 61:197-202.

Ferris, C. R. 1979. Effects of Interstate 95 on Breeding Birds in Northern Maine. *Journal of Wildlife Management* 43:421-27.

Findlay, C. S., and J. Bourdages. 2000. Response Time of Wetland Biodiversity to Road Construction on Adjacent Lands. *Conservation Biology* 14:86-94.

Findlay, C. S., and J. Houlahan. 1997. Anthropogenic Correlates of Species Richness in Southeastern Ontario Wetlands. *Conservation Biology* 11:1000-09.

Foppen, R., and R. Reijnen. 1994. The Effects of Car Traffic on Breeding Bird Populations in Woodland. II. Breeding Dispersal of Male Willow Warblers (*Phylloscopus Trochilus*) in Relation to the Proximity of a Highway. *Journal of Applied Ecology* 31:95-101.

Forman, R. T. T., and R. D. Deblinger. 2000. The Ecological Road-Effect Zone of a Massachusetts (U.S.A.) Suburban Highway. *Conservation Biology* 14:36

Forman, R. T. T., B. Reineking, and A. M. Hersperger. 2002. Road Traffic and Nearby Grassland Bird Patterns in a Suburbanizing Landscape. *Environmental Management* 29:782-800.

Forman, R. T. T., and et al. 2003. *Road Ecology: Science and Solutions.* Island Press, Washington, D.C.

Forysa, E. A., C. R. Allen, and D. P. Wojcikc. 2002. Influence of the Proximity and Amount of Human Development and Roads on the Occurrence of the Red Imported Fire Ant in the Lower Florida Keys. *Biological Conservation* 108:27-33.

Fowle, S. C. 1990. *The Painted Turtle in the Mission Valley of Western Montana.* Master's. University of Montana, Missoula.

Fuller, T. 1989. Population Dynamics of Wolves in North-Central Minnesota. *Wildlife Monographs* 105:1-41.

Garland, T. J., and W. G. Bradley. 1984. Effects of Highway on Mojave Desert Rodent Populations. *American Midland Naturalist* 111:47-56.

Gerlach, G., and K. Musolf. 2000. Fragmentation of Landscape as a Cause for Genetic Subdivision in Bank Voles. *Conservation Biology* 14:1066-74.

Getz, L. L., F. R. Cole, and D. L. Gates. 1978. Interstate Roadsides as Dispersal Routes for Microtus Pennsylvanicus. *Journal of Mammalogy* 59:208-12.

Gibbs, J. P. 1998. Amphibian Movements in Response to Forest Edges, Roads, and Streambeds in Southern New England. *Journal of Wildlife Management* 62:584-89.

Gibeau, M. L. 2000. *A Conservation Biology Approach to Management of Grizzly Bears in Banff National Park, Alberta.* Ph.D. University of Calgary, Alberta.

Gilbert, O. L. 1989. *The Ecology of Urban Habitats.* Chapman and Hall Ltd. New York.

Goldsmith, C. D., and P. F. Scanlon. 1977. Lead Levels in Small Mammals and Selected Invertebrates Associated with Highways of Different Traffic Densities. *Bulletin of Environmental Contamination and Toxicology* 17:311-16.

Goodwin, C. R. 1987. *Tidal-Flow, Circulation, and Flushing Changes Caused by Dredge and Fill in Tampa Bay, Florida.* U.S.G.S. Water-Supply Paper 2282, U.S. Geological Survey, Washington, D.C.

Goosem, M. 2000. Effects of Tropical Rainforest Roads on Small Mammals: Edge Changes in Community Composition. *Wildlife Research* 27:151-63.

Goosem, M. 2001. Effects of Tropical Rainforest Roads on Small Mammals: Inhibition of Crossing Movements. *Wildlife Research* 28:351-64.

Greenberg, C. H., S. H. Crownover, and D. R. Gordon. 1997. Roadside Soils: A Corridor for Invasion of Xeric Scrub by Nonindigenous Plants. *Natural Areas Journal* 17:99-109.

Grue, C. E., D. J. Hoffman, W. N. Beyer, and L. P. Franson. 1986. Lead Concentrations and Reproductive Success in European Starlings (*Sturnus Vulgaris*) Nesting within Highway Roadside Verges. *Environmental Pollution (Series A)* 42:157-82.

Haskell, D. G. 2000. Effects of Forest Roads on Macroinvertebrate Soil Fauna of the Southern Appalachian Mountains. *Conservation Biology* 14:57.

Havlick, D. G. 2002. *No Place Distant*. Island Press, Washington, D.C.

Hels, T., and E. Buchwald. 2001. The Effect of Road Kills on Amphibian Populations. *Biological Conservation* 99:331-40.

Huey, L. M. 1941. Mammalian Invasion Via the Highway. *Journal of Mammalogy* 22:383-85.

Huijser, M. P., and P. J. M. Bergers. 2000. The Effect of Roads and Traffic on Hedgehog (*Erinaceus Europaeus*) Populations. *Biological Conservation* 95:111-16.

Jaeger, J. A. G., and L. Fahrig. 2001. Modeling the Effects of Road Network Patterns on Population Persistence: Relative Importance of Traffic Mortality and 'Fence Effect.' Pages 298-312 in G. Evink et al., eds. Proceedings of the International Conference on Ecology and Transportation. Centre for Transportation and the Environment, North Carolina State University, Raleigh, North Carolina USA.

Jensen, W. F., T. K. Fuller, and W. L. Robinson. 1986. Wolf (*Canis Lupus*), Distribution on the Ontario-Michigan Border near Sault Ste. Marie. *Canadian Field Naturalist* 100:363-66.

Jones, J. A., F. J. Swanson, B. C. Wemple, and K. U. Snyder. 2000. Effects of Roads on Hydrology, Geomorphology, and Disturbance Patches in Stream Networks. *Conservation Biology* 14:76-85.

Jones, M. E. 2000. Road Upgrade, Road Mortality and Remedial Measures: Impacts on a Population of Eastern Quolls and Tasmanian Devils. *Wildlife Research* 27:289-96.

Kasworm, W. F., and T. Manley. 1990. Road and Trail Influences on Grizzly and Black Bears in Northwest Montana. *International Conference on Bear Research and Managemet* 8:79-84.

Keller, V. E. 1991. The Effect of Disturbance from Roads on the Distribution of Feeding Sites of Geese (*Anser Brachyrhynchus, A. Anser*), Wintering in North-East Scotland. *Ardea* 79:229-32.

Kiang, Y. T. 1982. Local Differentiation of *Anthoxanthum Odoratum* L. Populations on Roadsides. *American Midland Naturalist* 107:340-50.

Kjensmo, J. 1997. The Influence of Road Salts on the Salinity and the Meromictic Stability of Lake Svinsjoen, Southern Eastern Norway. *Hydrobiologia*:151-58.

Knight, R. L., and J. Y. Kawashima. 1993. Responses of Raven and Red-Tailed Hawk Populations to Linear Right-of-Ways. *Journal of Wildlife Management* 57:266-71.

Knight, R. L., H. A. L. Knight, and R. J. Camp. 1995. Common Ravens and Number and Type of Linear Rights-of-Way. *Biological Conservation* 74:65-67.

Kobringer, N. P. 1984. *Sources and Migration of Highway Runoff Pollutants-Executive Summary. Vol. 1.* FHWA/RD-84/057, Federal Highway Administration and Rexnord EnviroEnergy Technology Center, Milwaukee, Wis.

Koford, E. J. 1993. Assessment and Mitigation for Endangered Vernal Pool Invertebrates. Pages 839-41 *in Conference Proceedings for the 20th Anniversary Conference on Water Management in the 90's.* Water Resource Planning and Management of Urban Water Resources, ASCE, New York.

Kozel, R. M., and E. D. Fleharty. 1979. Movement of Rodents across Roads. *Southwestern Naturalist* 24:239-48.

La March, J. L., and D. P. Lettenmaier. 2001. Effects of Forest Roads on Flood Flows in Deschutes River, Washington. *Earth Surface Processes and Landforms* 26:115-34.

Likens, G. E., F. H. Boarmann, R. S. Pierce, J. S. Eaton, and N. M. Johnson. 1977. *Biogeochemistry of a Forest Ecosystem.* Springer-Verlag, New York.

Lonsdale, W. M., and A. M. Lane. 1994. Tourist Vehicles as Vectors of Weed Seeds in Kakadu National Park, Northern Australia. *Biological Conservation* 69:277-83.

Lovallo, M. J., and E. M. Anderson. 1996. Bobcat Movements and Home Ranges Relative to Roads in Wisconsin. *Wildlife Society Bulletin* 24:71-76.

Loving, B. L., K. M. Waddell, and C. W. Miller. 2000. *Water and Salt Balance of Great Salt Lake, Utah, and Simulation of Water and Salt Movement through the Causeway, 1987-98.* U.S.G.S. Water Resources Investigation Report 00-4221, U.S. Geological Survey, Salt Lake City, Utah.

Lyon, L. J. 1983. Road Density Models Describing Habitat Effectiveness for Elk. *Journal of Forestry* 81:592-95.

Lytle, C. M., B. N. Smith, and C. Z. Mckinnon. 1995. Manganese Accumulation Along Utah Roadways: A Possible Indication of Motor Vehicle Exhaust Pollution. *Science and the Total Environment* 162:105-9.

Mace, R. D., J. S. Waller, T. L. Manley, L. J. Lyon, and H. Zuuring. 1996. Relationships among Grizzly Bears, Roads and Habitat in the Swan Mountains, Montana. *Journal of Applied Ecology* 33:1395-404.

Madej, M. A. 2001. Erosion and Sediment Delivery Following Removal of Forest Roads. *Earth Surface Processes and Landforms* 26:175-90.

Mader, H. J. 1984. Animal Habitat Isolation by Roads and Agricultural Fields. *Biological Conservation* 29:81-96.

Mader, H. J., C. Schell, and P. Kornacker. 1990. Linear Barriers to Arthropod Movements in the Landscape. *Biological Conservation* 54:209-22.

Mahaney, P. A. 1994. Effects of Freshwater Petroleum Contamination on Amphibian Hatching and Metamorphosis. *Environmental Toxicology* 13:259-65.

Marino, F., A. Ligero, and D. J. Diaz Cosin. 1992. Heavy Metals and Earthworms on the Border of a Road Next to Santiago (Galicia, Northwest of Spain): Initial Results. *Soil Biology and Biochemistry* 24:1705-9.

Massachusetts EOEA (Executive Office of Environmental Affairs). 1995. *Phragmites- Controlling the All-Too-Common Reed.* Commonwealth of Massachusetts, Boston.

Mech, L. D., S. H. Fritts, G. L. Radde, and W. J. Paul. 1988. Wolf Distribution and Road Density in Minnesota. *Wildlife Society Bulletin* 16:85-87.

Megahan, W. F. 1980. *Effects of Silvicultural Practices on Erosion and Sedimentation in the Interior West: A Case for Sediment Budgeting.* U.S. Department of Agricultural, Forest Service, Intermountain Forest and Range Experiment Station, Boise, ID.

Merriam, G., K. Michal, E. Tsuchiya, and K. Hawley. 1989. Barriers as Boundaries for Metapopulations and Demes of *Peromyscus Leucopus* in Farm Landscapes. *Landscape Ecology* 29:227-35.

Merrill, S. B. 2000. Road Densities and Wolf, *Canis Lupus*, Habitat Suitability; an Exception. *Canadian Field Naturalist* 114:312-13.

Meunier, F. D., C. Verheyden, and P. Jouventin. 2000. Use of Roadsides by Diurnal Raptors in Agricultural Landscapes. *Biological Conservation* 92:291-98.

Miller, J. R., L. A. Joyce, R. L. Knight, and R. M. King. 1996. Forest Roads and Landscape Structure in the Southern Rocky Mountains. *Landscape Ecology* 11:115-27.

Ministry of Transport, P. W. A. W. M. 1994a. *Towards Sustainable Verge Management in the Netherlands.* No. 59, Ministerie van Verkaar en Waterstaat, Delft, Netherlands.

Ministry of Transport, P. W. A. W. M. 1994b. *The Chemical Quality of Verge Grass in the Netherlands.* No. 62, Dienst Weg- en Waterbouwkunde, Ministerie van Verkaar en Waterstaat, Delft, Netherlands.

Ministry of Transport, P. W. A. W. M. 2000. *National Highway Verges...National Treasures!* Ministerie van Verkaar en Waterstaat, Delft, Netherlands.

Minoransikii, V. A., and K. Z. R. 1984. Effect of Environmental Pollution by Motor Transport on the Reproduction and Development of *Opatrum Sabulosum. Biologicheskie Nauki (Moscow)* 0:43-7.

Mladenoff, D. J., T. A. Sickley, R. G. Haight, and A. P. Wydeven. 1995. A Regional Landscape Analysis of Favorable Gray Wolf Habitat in the Northern Great Lakes Region. *Conservation Biology* 9:279-94.

Mladenoff, D. J., T. A. Sickley, and A. P. Wydeven. 1999. Predicting Gray Wolf Landscape Recolonization: Logistic Regression Models vs. New Field Data. *Ecological Applications* 9:37-44.

Mumme, R. L., S. J. Schoech, G. E. Woolfenden, and J. W. Fitzpatrick. 2000. Life and Death in the Fast Lane: Demographic Consequences of Road Mortality in the Florida Scrub-Jay. *Conservation Biology* 14:501-12.

Munguira, M. L., and J. A. Thomas. 1992. Use of Road Verges by Butterfly and Burnet Populations, and the Effect of Roads on Adult Dispersal and Mortality. *Journal of Applied Ecology* 29:316-29.

Murphy, E. C., and J. E. Dowding. 1994. Range and Diet of Stoats (*Mustela Erminea*) in a New Zealand Beech Forest. *New Zealand Journal of Ecology* 19:11-18.

Muskett, C. J., and M. P. Jones. 1980. The Dispersal of Lead, Cadmium and Nickel from Motor Vehicles and Effects on Roadside Invertebrate Macrofauna. *Environmental Pollution (Series A)* 23:231-42.

Nellemann, C., and R. D. Cameron. 1996. Effects of Petroleum Development on Terrain Preferences of Calving Caribou. *Arctic* 49:23-28.

Nellemann, C., and R. D. Cameron. 1998. Cumulative Impacts of an Evolving Oil-Field Complex on the Distribution of Calving Caribou. *Canadian Journal of Zoology* 76:1425-30.

Nellemanna, C., I. Vistnesb, P. Jordhøyc, and O. Strandc. 2001. Winter Distribution of Wild Reindeer in Relation to Power Lines, Roads and Resorts. *Biological Conservation* 101:351-60.

Ortega, Y. K., and D. E. Capen. 1999. Effects of Forest Roads on Habitat Quality for Ovenbirds in a Forested Landscape. *Auk* 116:937-46.

Oxley, D. J., M. B. Fenton, and G. R. Carmody. 1974. The Effects of Roads on Small Mammals. *Journal of Applied Ecology* 11:51-59.

Parendes, L. A., and J. A. Jones. 2000. Role of Light Availability and Dispersal in Exotic Plant Invasion Along Roads and Streams in the H. J. Andrews Experimental Forest, Oregon. *Conservation Biology* 14:64-75.

Port, G. R., and J. R. Thompson. 1980. Outbreaks of Insect Herbivores on Plants Along Motorways in the United Kingdom. *Journal of Applied Ecology* 17:649-56.

Przybylski, Z. 1979. The Effects of Automobile Exhaust Gases on the Arthropods of Cultivated Plants, Meadows and Orchards. *Environmental Pollution* 19:937-49.

Reed, R. A., J. Johnson-Barnard, and W. L. Baker. 1996. Contribution of Roads to Forest Fragmentation in the Rocky Mountains. *Conservation Biology* 10:1098-106.

Reh, W., and A. Seitz. 1990. The Influence of Land Use on the Genetic Structure of Populations of the Common Frog *Rana Temporaria*. *Biological Conservation* 54:239-49.

Reijnen, R., and R. Foppen. 1994. The Effects of Car Traffic on Breeding Bird Populations in Woodland. I. Evidence of Reduced Habitat Quality for Willow Warblers (*Phylloscopus Trochilus*) Breeding Close to a Highway. *Journal of Applied Ecology* 31:85-94.

Reijnen, R., and R. Foppen. 1995. The Effects of Car Traffic on Breeding Bird Populations in Woodland. IV. Influence of Population Size on the Reduction of Density Close to a Highway. *Journal of Applied Ecology* 32:481-91.

Reijnen, R., R. Foppen, C. Ter Braak, and J. Thissen. 1995. The Effects of Car Traffic on Breeding Bird Populations in Woodland. III. Reduction of Density in Relation to the Proximity of Main Roads. *Journal of Applied Ecology* 32:187-202.

Reijnen, R., R. Foppen, and H. Meeuwsen. 1996. The Effects of Traffic on the Density of Breeding Birds in Dutch Agricultural Grasslands. *Biological Conservation* 75:255-60.

Reijnen, R., R. Foppen, and G. Veenbaas. 1997. Disturbance by Traffic of Breeding Birds: Evaluation of the Effect and Considerations in Planning and Managing Road Corridors. *Biodiversity and Conservation* 6:567-81.

Rich, A. C., D. S. Dobkin, and L. J. Niles. 1994. Defining Forest Fragmentation by Corridor Width: The Influence of Narrow Forest-Dividing Corridors on Forest-Nesting Birds in Southern New Jersey. *Conservation Biology* 8:1109-12.

Richardson, J. H., R. F. Shore, and J. R. Treweek. 1997. Are Major Roads a Barrier to Small Mammals? *Journal of Zoology* 243:840-46.

Ries, L., D. M. Debinski, and M. L. Wieland. 2001. Conservation Value of Roadside Prairie Restoration to Butterfly Communities. *Conservation Biology* 15:401-11.

Roach, G., and R. Kirkpatrick. 1985. Wildlife Use of Roadside Woody Plantings in Indiana. *Transportation Research Record* 1016:11-15.

Rosen, P. C., and C. H. Lowe. 1994. Highway Mortality of Snakes in the Sonoran Desert of Southern Arizona. *Biological Conservation* 68:143-48.

Rost, G. R., and J. A. Bailey. 1979. Distribution of Mule Deer and Elk in Relation to Roads. *Journal of Wildlife Management* 43:634-41.

Saunders, S. C., M. R. Mislivets, J. Chen, and D. T. Cleland. 2002. Effects of Roads on Landscape Structure within Nested Ecological Units of the Northern Great Lakes Region, USA. *Biological Conservation* 103:209-25.

Schloss, J. A. 2002. GIS Watershed Mapping: Developing and Implementing a Watershed Natural Resources Inventory (New Hampshire) *in* R. L. France, editor. *Handbook of Water Sensitive Planning and Design*. Lewis Publishers, Boca Raton, Fla.

Schueler, T. 1995. *Site Planning for Urban Stream Protection.* Center for Watershed Protection, Ellicot City, Md.

Spencer, H. J., and G. R. Port. 1988. Effects of Roadside Conditions on Plants and Insects. II. Soil Conditions. *Journal of Applied Ecology* 25:709-15.

Stiles, E. W. 1980. Patterns of Fruit Presentation and Seed Dispersal in Bird-Disseminated Woody Plants in the Eastern Deciduous Forest. *American Midland Naturalist* 116:670-88.

Swanson, F. J., and C. T. Dyrness. 1975. Impact of Clearcutting and Road Construction on Soil Erosion by Landslides in the Western Cascade Range, Oregon. *Geology* 3:393-96.

Swihart, R. K., and N. A. Slade. 1984. Road Crossing in *Sigmodon Hispidus* and *Microtus Ochrogaster*. *Journal of Mammalogy* 65:357-60.

Sykora, K. V., L. J. D. Nijs, and T. A. H. M. Pelsma. 1993. *Plantengemeenschappen Van Nederlandse Wegbermen.* Stichting Uitgeverij Koninklijke Nederlandse Natuurhistorische Vereniging, Utrecht, Netherlands.

Theil, R. P. 1985. Relationship between Road Densities and Wolf Habitat Suitability in Wisconsin. *American Midland Naturalist* 113:404-07.

Thompson, J. R., and A. J. Rutter. 1986. The Salinity of Motorway Soils: IV. Effects of Sodium Chloride on Some Native British Shrubs Species, and the Possibility of Establishing Shrubs on the Central Reserves of Motorways. *Journal of Applied Ecology* 23:299-315.

Thurber, J. M., R. O. Peterson, T. D. Drummer, and S. A. Thomasma. 1994. Gray Wolf Response to Refuge Boundaries and Roads in Alaska. *Wildlife Society Bulletin* 22:61-68.

Tikka, P. M., H. Hogmander, and P. S. Koski. 2001. Road and Railway Verges Serve as Dispersal Corridors for Grassland Plants. *Landscape Ecology* 16:659-66.

Turtle, S. L. 2000. Embryonic Survivorship of the Spotted Salamander (*Ambystoma Maculatum*) in Roadside and Woodland Vernal Pools in Southeastern New Hampshire. *Journal of Herpetology* 34:60-67.

Udevitz, M. S., C. A. Howard, R. J. Robel, and B. Curnutte. 1980. Lead Contamination in Insects and Birds near an Interstate Highway. *Environmental Entomology* 9:35-36.

Ullmann, I., P. Bannister, and J. B. Wilson. 1998. The Vegetation of Roadside Verges with Respect to Environmental Gradients in Southern New Zealand. *Journal of Vegetation Science* 6:131-42.

van der Zande, A. N., W. J. Ter Keurs, and W. J. Van Der Weijden. 1980. The Impact of Roads on the Densities of Four Bird Species in an Open Field Habitat—Evidence of a Long-Distance Effect. *Biological Conservation* 18:299-321.

van der Zee, F. F., J. Wiertz, J. F. Ter Braak, and R. C. Van Apeldoorn. 1992. Landscape Change as a Possible Cause of the Badger *Meles Meles* L. Decline in the Netherlands. *Biological Conservation* 61:17-22.

Vermeulen, H. J. W. 1994. Corridor Function of a Road Verge for Dispersal of Stenotopic Heathland Ground Beetles (*Carabidae*). *Biological Conservation* 69:339-49.

Vos, C. C., and J. P. Chardon. 1998. Effects of Habitat Fragmentation and Road Density on the Distribution Pattern of the Moor Frog *Rana Arvalis*. *Journal of Applied Ecology* 35:44-56.

Wade, K. J., J. T. Flanagan, A. Currie, and D. J. Curtis. 1980. Roadside Gradients of Lead and Zinc Concentrations in Surface-Dwelling Invertebrates. *Environmental Pollution (Series B)* 1:87-93.

Walker, D. A., D. Cate, J. Brown, and C. Racine. 1987. *Disturbance and Recovery of Artic Alaskan Tundra Terrain: A Review of Recent Investigations.* Report 87-11, Cold Regions Research and Engineering Laboratory, Hanover, N.H.

Walker, D. A., P. J. Webber, E. F. Binnian, K. R. Everett, N. D. Lederer, E. A. Norstrand, and M. D. Walker. 1987. Cumulative Impacts of Oil Fields on Northern Alaskan Landscapes. *Science* 338:757-61.

Warner, R. E. 1992. *Nest Ecology of Grassland Passerines on Road Rights-of-Way in Central Illinois (Revised).* PB96-116330, National Technical Information Service.

Warren, M. L., and M. G. Pardew. 1998. Road Crossings as Barriers to Small-Stream Fish Movement. Trans. Amer. Fish Soc. 127, 637-644. *Transactions of the American Fisheries Society* 127:637-44.

Watkins, L. H. 1981. *Environmental Impact of Roads and Traffic.* Applied Science Publishers, London.

Way, J. M. 1977. Roadside Verges and Conservation in Britain: A Review. Biological Conservation. 12: 65-74.

Wemple, B. C., J. A. Jones, and G. E. Grant. 1996. Channel Network Extension by Logging Roads in Two Basins, Western Cascades, Oregon. *Water Resources Bulletin* 32:1195-207.

Wemple, B. C., F. J. Swanson, and J. A. Jones. 2001. Forest Roads and Geomorphic Process Interactions, Cascade Range, Oregon. *Earth Surface Processes and Landforms* 26:191-204.

Wilcox, D. A. 1989. Migration and Control of Purple Loosestrife (*Lythrum Salicaria* L.) Along Highway Corridors. *Environmental Management* 13: 365-70.

Wilkins, K. T. 1982. Highways as Barriers to Rodent Dispersal. *Southwestern Naturalist* 27:459-60.

Williamson, P., and P. R. Evans. 1972. Lead: Levels in Roadside Invertebrates and Small Mammals. *Bulletin of Environmental Contamination and Toxicology* 8:280-88.

Wong, M. H., Y. H. Cheung, and W. W. C. 1984. Effects of Roadside Germination and Root Growth of *Brassica Chinensis* and *B. Parachinensis*. *The Science of the Total Environment* 33:87-102.

Woodard, S. M. 1990. Population Density and Home Range Characteristics of Woodchucks, *Marmota Monax*, at Expressway Interchanges. *Canadian Field-Naturalist* 104:421-28.

Zeigler, A. D., R. A. Sutherland, and T. W. Giambelluca. 2001. Interstorm Surface Preparation and Sediment Detachment by Vehicle Traffic on Unpaved Mountain Roads. *Earth Surface Processes and Landforms* 26:235-50.

# Appendix C

# Congressional Declaration of
# National Environmental Policy
## National Environmental Policy Act of 1969

### UNITED STATES CODE, TITLE 42, SECTION 4331

(a) The Congress, recognizing the profound impact of man's activity on the interrelations of all components of the natural environment, particularly the profound influences of population growth, high-density urbanization, industrial expansion, resource exploitation, and new and expanding technological advances and recognizing further the critical importance of restoring and maintaining environmental quality to the overall welfare and development of man, declares that it is the continuing policy of the Federal Government, in cooperation with State and local governments, and other concerned public and private organizations, to use all practicable means and measures, including financial and technical assistance, in a manner calculated to foster and promote the general welfare, to create and maintain conditions under which man and nature can exist in productive harmony, and fulfill the social, economic, and other requirements of present and future generations of Americans.

(b) In order to carry out the policy set forth in this Act, it is the continuing responsibility of the Federal Government to use all practicable means, consistent with other essential considerations of national policy, to improve and coordinate Federal plans, functions, programs, and resources to the end that the Nation may—

1.  fulfill the responsibilities of each generation as trustee of the environment for succeeding generations;

2.    assure for all Americans safe, healthful, productive, and aesthetically and culturally pleasing surroundings;

3.    attain the widest range of beneficial uses of the environment without degradation, risk to health or safety, or other undesirable and unintended consequences;

4.    preserve important historic, cultural, and natural aspects of our national heritage, and maintain, wherever possible, an environment which supports diversity, and variety of individual choice;

5.    achieve a balance between population and resource use which will permit high standards of living and a wide sharing of life's amenities; and

6.    enhance the quality of renewable resources and approach the maximum attainable recycling of depletable resources.

(c) The Congress recognizes that each person should enjoy a healthful environment and that each person has a responsibility to contribute to the preservation and enhancement of the environment.

Information available at: http://ceq.eh.doe.gov/nepa/nepanet.htm (accessed March 2, 2005).